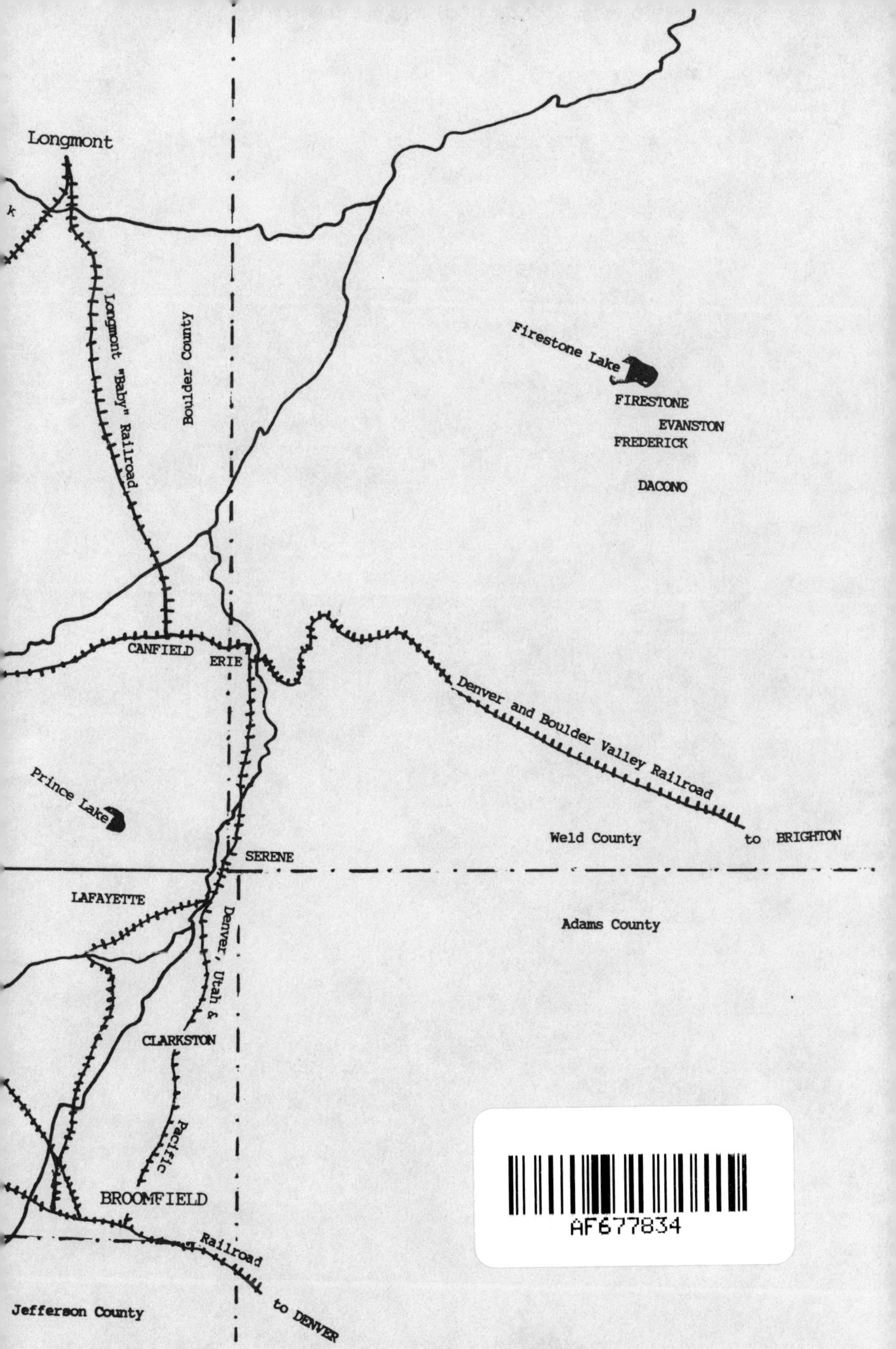
Longmont
Longmont "Baby" Railroad
Boulder County
Firestone Lake
FIRESTONE
EVANSTON
FREDERICK
DACONO
CANFIELD
ERIE
Denver and Boulder Valley Railroad
Prince Lake
Weld County
to BRIGHTON
SERENE
LAFAYETTE
Adams County
Denver, Utah &
CLARKSTON
Pacific
BROOMFIELD
Railroad
to DENVER
Jefferson County

ONCE A COAL MINER...

ONCE A COAL MINER

The Story of Colorado's Northern Coal Field

By
Phyllis Smith

PRUETT PUBLISHING COMPANY
Boulder, Colorado

First Edition

1 2 3 4 5 6 7 8 9

Library of Congress Cataloging-in-Publication Data

Smith, Phyllis, 1928—
Once a coal miner: the story of Colorado's northern coal fields/by Phyllis Smith.—1st ed.
p. cm.
Bibliography: p.
Includes index.
ISBN 0-87108-774-X
1. Coal miners—Colorado—History. I. Title.
HD8039.M62ZU66838 1989
338.7′622334′09788—dc20 89-33684
CIP

A last remnant of the Northern Coal Field.—*Photo,* Longmont Times-Call

Once you're a coal miner, you don't want to do anything else.

"Red" Schofield, Lafayette miner

Coal miners take a break and pose for the camera.—*Photo, Louisville Public Library*

Coal mining was the only job I ever had that I didn't mind getting up in the morning.

Jimmie W. Graham, Jr., Lafayette

Erecting timbers for underground safety.—*Photo, Lafayette Public Library*

Once you're a coal miner, you never give it up. It gets in your blood, like dope, and you can never leave it. I've been on strikes where I'd get mad and go find another job just as good or better, but when snow flakes started flying, why, I wanted to be underground. That's a coal miner. He's just a breed of his own.

Lawrence Johnson

Langford coal miners pose for the camera of Joseph B. Sturtevant.—*Photo, A.A. Paddock Collection, Boulder Historical Society, Carnegie Branch Library for Local History*

Where else can you work where it's warm in the winter and cool in the summer?

John Hutchison, Lafayette miner

Winter in the Northern Coal Field.—*Photo, Boulder Historical Society, Carnegie Branch Library for Local History*

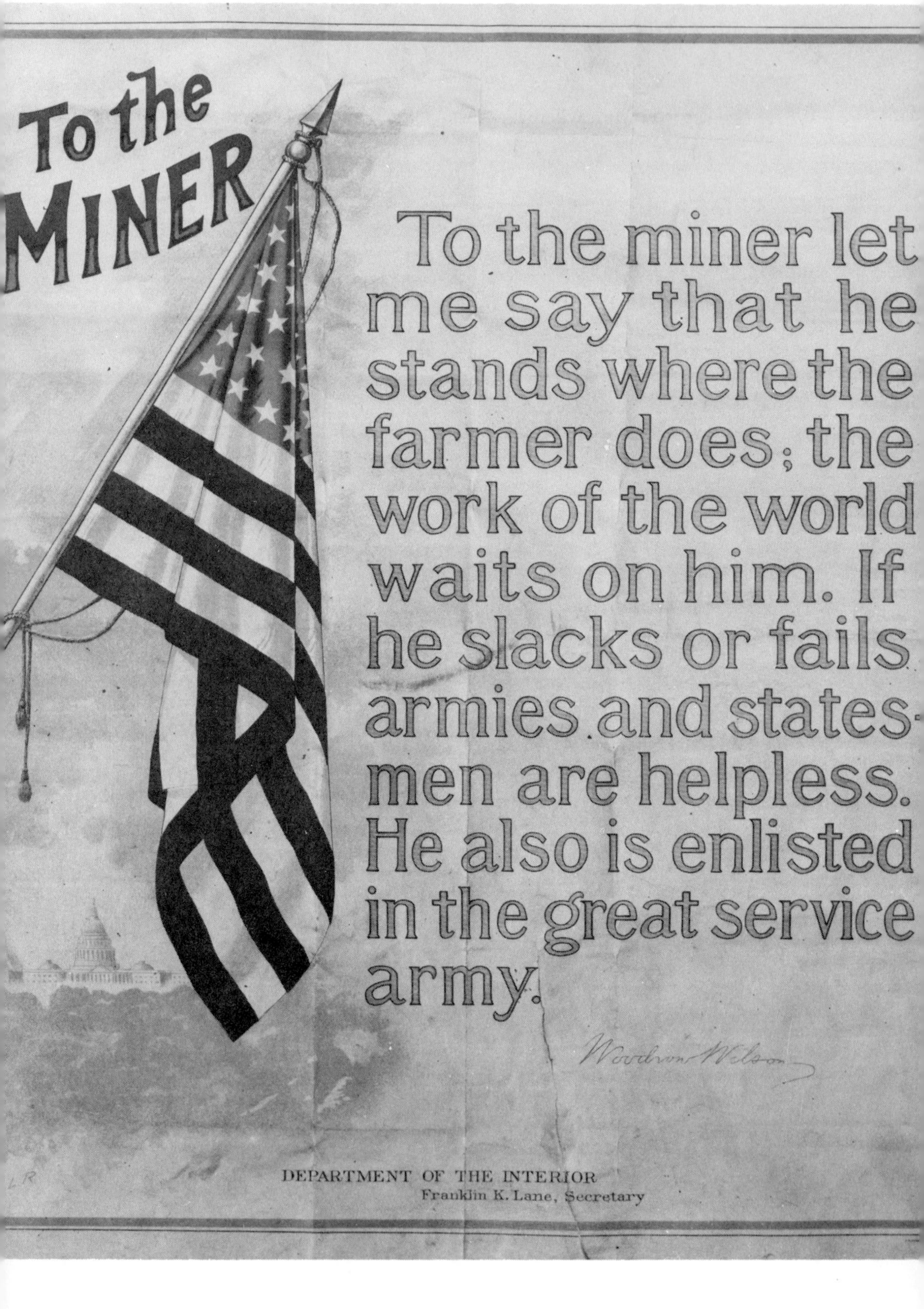
To the MINER
To the miner let me say that he stands where the farmer does; the work of the world waits on him. If he slacks or fails armies and statesmen are helpless. He also is enlisted in the great service army.
Woodrow Wilson
DEPARTMENT OF THE INTERIOR
Franklin K. Lane, Secretary

The men at the Lehigh mine, 1903.—*Photo, Daisy Bracegirdle Swallow*

CONTENTS

Coal miners in front of a building. Note lard oil lamps on caps and graniteware lunch pail.
—Photo, Western Historical Collections, University of Colorado at Boulder

ACKNOWLEDGMENTS

No one writes history in a vacuum. Since none of my people were coal miners, nor did I grow up in the Northern Coal Field, I have depended upon the thoughts of those who were kind enough to visit with me, or who were far-sighted enough to set down their reminiscences on paper or on tape. Those northern field residents who helped me are listed at the end of this section.

The single most important source for this subject is the 1959 doctoral dissertation of Donald J. McClurg, a well-organized and thoughtful work.

Two friends put up with alternate moods of exhilaration and despair: Donna Hudgel, proprietor of Trails West, an unusual bookstore based in Louisville specializing in western Americana, and Virginia Braddock, municipal reference librarian at the Boulder Public Library. Both were endlessly patient with me.

Members of the Boulder Writers Club gave valuable suggestions as they listened to the completed manuscript.

Photographer Don Look of Boulder, who likes old photographs, worked with a loving hand on most of the old pictures.

In Lafayette, I am indebted to librarians Blanche Moon and LaJuan Swallow, former librarian Effie Amicarella, and museum director Ruthene Rodwick for the kind of wisdom that can come only from people whose families settled the area generations before. Long-time residents Jim and Beth Hutchison and Winston Morgan were also gracious.

In Louisville, former librarian Ann Tremontozzi opened her historical files to me; volunteer Vera Taylor, descendant of one of the first Superior families, helped me to identify photographs.

Former Erie teacher Irma Bledsoe and present teacher Karen Adelfang were very helpful. W.A. Warnock of Loveland, whose family settled in early Erie, contributed photographs from the family album. Steve Bednar of Superior was supportive.

Laurence Paddock, editor of the Boulder *Daily Camera*, has been an intelligent source of information; I am grateful that three generations of Paddocks have been pack rats, collecting photographs and every bit of memorabilia that came their way. *Camera* librarian Charlotte Smokler has always been available to help me search out information.

Cassandra Volpe of the Western Historical Collections, University of Colorado at Boulder, has been resourceful and helpful to me. Lecturer Jim Whiteside, University of Colorado, pointed the way to fresh material; Professor Howard Lee Scamehorn, also of the University of Colorado and an expert on the Colorado Fuel and Iron Company, has played devil's advocate. Eric Margolis, formerly of the Coal Project, was an enthusiastic listener.

At the Carnegie Branch Library for Local History, Boulder, librarian archivist Lois Anderton extended herself to assist me with research possibilities; photographic volunteer Kay Stiewig unearthed a number of fine pictures. The Boulder community is fortunate to have such outstanding reference librarians at the Boulder Public Library: Barbara Buchman, Jane Perlmutter, Lynn Reed, Connie Walker, Judy Waller, and Lu Wright. Boulder residents Silvia Pettem, Franklin Folsom, Rich Koopman, and Sender Garlin also furnished me with information.

Eric Paddock, archivist at the Colorado Historical Society, was able to come up with a number of important photographs. Augie Mastroguiseppe and Kathey Swan of the Western History Department, Denver Public Library, were helpful as well.

The following people have contributed their memories of earlier times either by personal interview, written statement, or taped recording as part of an oral history project:

Cliff Alderson
Claude Amicarella
Effie Amicarella
Lawrence Amicarella
Anna Barrowman
William Barrowman
Andy Borrego
Mary Borstnick
Mike Brennan
Sarah Savage Brillhart
Kenneth Brown
Frank Brugger
Jack Davis
Ruth Davis
Ralph DeAndrea
Dorothy Eastenas
W.B. Edwards
Hannah Evans
Steve Fotis
M.P. Fox
Steve Gorce
Bessie Graham
Jimmie Graham, Jr.
Jack Green
Ann Guerreri
Jessie Irwin James
John James
Robert Johnson
George Jones
Thomas Kerr
Elizabeth Lewis
Elmo Lewis
John "Greekie" Lewis
James Ravell Lord
George Malcoff
Frank "Ponzo" Mastriana
Henry "Welchy" Mathias
Joe Mathias
Frank Miller
Ralph Clinton Miller
Willeta Miller
Polly Simpson Morgan
John Murdock
Boughton "Boots" Noble
Clara Autrey O'Donnell
Angela "Pete" Quella
Bert Pierson
Joe Rodwick
Frank E. Sargent
Dugan "Red" Schofield
Mildred Green Sirokman
Harold Spicer
Daisy Bracegirdle Swallow
Clancy Waneka
Arlo Wilson
William O. Wise

INTRODUCTION

Come all you young fellows, so young and so fine,
And seek not your fortune in a dark dreary mine.
It will form as a habit and seep in your soul,
'Till the stream of your blood runs as black as the coal.

Where it's dark as a dungeon and damp as the dew,
Where the danger is double and the pleasures are few,
Where the rain never falls and the sun never shines,
It's dark as a dungeon way down in the mines.[1]

This is the story of the miners of the Northern Coal Field, a land that stretched across the rolling prairie just in sight of the Rocky Mountains of central Colorado.

The Northern Coal Field, which encompassed 6,800 square miles, was characterized as a "rainless, treeless waste"[2] whose broad-reaching skies were "sunny but unsympathetic and lonely."[3]

Most of the dark brown coal seams, lying beneath short, thick grass covering the plateau, were concentrated within a four-hundred-square-mile tract in eastern Boulder County and western Weld County. These fields of lignite and sub-bituminous coal formed some eighty to one hundred million years ago when the area was an inland sea and marsh. Some outcroppings from these coal banks lay exposed along the surface of the western ridges that tilted toward the foothills of the mountains just beyond.

Those who crossed the plains in the fall of 1858 and the following spring did not have coal on their minds, however. Instead, their eyes were fixed upon the Great Snowy Range, as the Rockies were called then, hoping for instant riches from the glamorous ultimate gold srike. They dreamed of glittering lumps of gold lying in heaps along the mountain streams beyond the foothills, waiting to be picked up by those who would get there first.

No doubt some of these would-be gold seekers stumbled and tripped over coal outcroppings as they crossed the ridges into the Boulder Valley and on to the mountains, where they planned to "git the gold and git out."

Most of these early travelers were modest farmers, small-town merchants, and unskilled day laborers, down on their luck. The panic of 1857 had contributed to hard times back in the States. Some of those moving west wanted to leave behind them the winds of an approaching civil war. Others, experiencing that special brand of American restlessness, came for the pure adventure of it, came "to see the elephant,"* and then went back home, satisfied that, at least for them, life here was no better than the one they left behind.

* "To see the elephant" is "to see the world and gain experience of its sin and glitter, generally at some cost to the investigator." Peter Watts, *A Dictionary of the Old West* (New York: Knopf, 1977), p. 125.

Some arrivals, however, regarded themselves as prospective settlers and looked over the territory rather carefully. They were tired of the dirt and growing congestion of the industrializing northeastern United States. They were brought down by recurrent illness in the family. They hoped their tubercular children would benefit by the clear, crystal air, the sparkling mountain water, and the prospect of a more vigorous life in a new land.

The winter weather was mild; the men could work outside in their shirtsleeves. They took note of the prairie grasses. Perhaps they would not have to slaughter their oxen and horses after all. The bands of Southern Arapaho Indians they encountered did not seem vicious or warlike. Some farmers were glad they had brought along seed for wheat and turnips. They decided to send for their families.

Just as news spread quickly in the early spring of 1859 about gold strikes in the mountains, so did word get back to coal miners and businessmen in the States of coal deposits lying in the tilted land below the Front Range. It was not long before the mining of coal beds in the Northern Coal Field directed, even consumed, the lives of the settlers and those who followed soon after. (See Appendix B for a discussion of the geology of the region.)

Along with the inexperienced adventurer came knowledgeable coal miners from the States as well as immigrant miners from England, Scotland, Wales, Ireland, and Germany. Many brought with them strongly-held views on the rights of labor—a gospel of unionism. These views branded the first miners as troublemakers almost immediately and led to the first confrontation between labor and management in Colorado a few years later.

When the introduction of railroad lines into the coal fields attracted eastern business interests a decade later, immigrants from France, Belgium, Poland, Italy, Croatia, Greece, and Russia were imported to work in the newly opened coal mines—more often than not as scab* labor.

These new arrivals seldom spoke English, knew little about coal mining at first, and were not well received by hostile Americans who feared their growing immigration would glut the labor market. However, the newcomers were filled with the promise of a new life, better than the one they left behind in the old country.

To the descendants of the first families who worked the mines of the Northern Coal Field, the violent days of their grandfathers' struggle against the coal operators are still vivid. The grandchildren know the old stories of the despised Baldwin-Felts men, imported from West Virginia by management to guard the mines and control the scabs. They have heard great-uncle tell tales about the state militia camped just outside of town.

Some know about the night bullets flew at the Hecla mine or the early-morning gunfire at the Columbine. They have been told about the federal troops sent in to stop the fighting. The young people also know that grandfather or great-uncle suffer now from black lung disease. Some can sing fragments of the old songs which, years ago, were vigorously sung in troubled times.

Is it possible to stand in downtown Louisville and hear again the 1903 words of

* The term "scab" refers to a worker introduced to the labor force to take the place of a worker on strike, generally regarded by strikers as the lowest form of life.

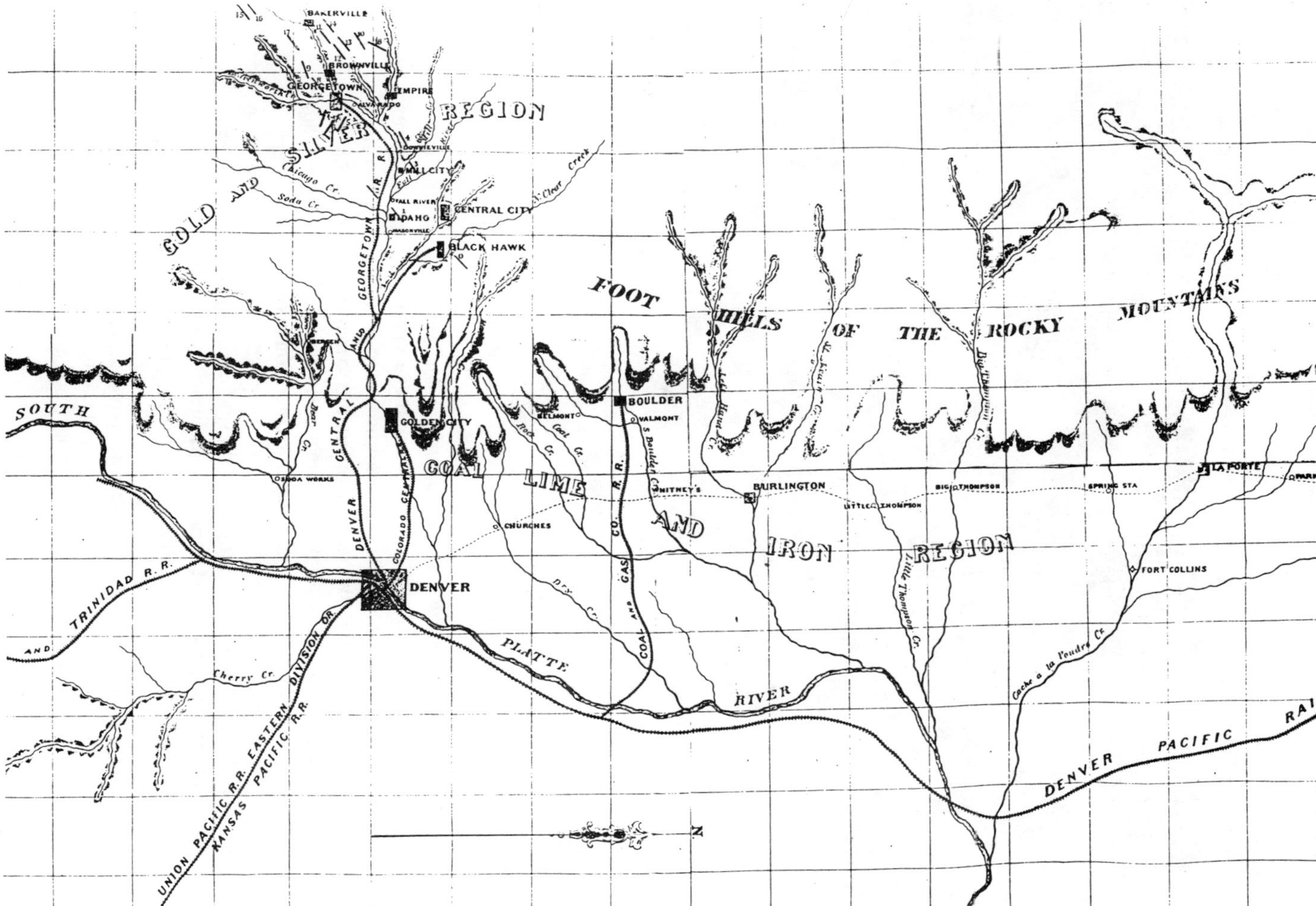

Undated map of the area, probably late 1860s; the artist anticipates the extension of the railroad into Boulder.—*Map, J.S. Randall Collection, Western Historical Collections,*

hell-raising Mother Jones:

> Are you brave men? Can you fight as well as you can work? I'd rather fall fighting than working. . . .The iron heel feels the same to all flesh. . . .We will all go to glory together or we will die and go down together.[4]

Standing near the site of the Columbine mine, perhaps we can imagine we hear the men and women singing as an early-morning trumpet calls them to the picket line. The first rays of the sun pick up the colors of little, fluttering American flags clutched by the miners' children.

The tipples* are gone now, but some folks can still see the guard sitting on top of the wooden tower, holding his machine gun at the ready, his circling searchlight picking up desperate figures running to get away from its dangerous beam. What sport it is for the laid-off miner, sitting on his front porch nearby with his rifle, to knock out the searchlight again and again.

The irrigation ditches are still there. Perhaps we can imagine we hear the

Two coal miners' wives rest a bit beside an irrigation ditch near Marshall before loading their buckets and heading for home.—*Photo, Boulder Historical Society, Carnegie Branch Library for Local History*

---

* A tipple is a wooden structure above ground that discharges coal from underground cars into a wagon, truck, or railroad car by means of a cable and pulley.

rushing water from one ditch, released to flush out the men crouched to avoid gunfire from the Hecla. Can we see the Greek labor organizer Louis Tikas dashing from a Lafayette boardinghouse, on the run from the Baldwin-Felts men?

Many of the immigrants imported to work as scabs joined the striking miners when they understood how they were being used. The man who persisted in scabbing was likely to be escorted by the striking miners to the edge of town, "shown the North Star and notified to travel until he came to it."[5]

The working miners seldom saw the day's sunlight except during the summer when the mines lay idle. At the end of a long shift of twelve or fourteen hours, they came up from underground to be greeted by their small sons, who grabbed the empty lunch pails and ran to the saloon to have them filled with their fathers' first beer, swallowed no doubt to rid themselves of the black, oily dust in their throats. A few of the women drank too, secretly, of course, and sent the young ones to the saloon's back door for a refill.

Tension in the coal camps was almost constant as the women listened for the long mine whistle that signaled an accident. If there was an explosion underground, they could feel the earth tremble beneath their feet before the whistle shrieked. They also listened, sometimes in vain, for the extra two whistle blasts that announced there would be work tomorrow.

From time to time, mining families were forced to trade in company stores, not

Louisville's "fire laddies" are ready to start the race between volunteer hose teams, 1880s.—*Photo, Louisville Public Library*

Lafayette's miners are ready to play ball, 1905.—*Photo, Lafayette Public Library*

Louisville's Rod and Gun Club.—*Ed Tangen photo, Louisville Public Library*

with real money but with scrip, issued by the coal operator as wages. If they were sick, they had to visit the company doctor, whose fees had already been taken out of the miners' wages.

In Colorado's Southern Coal Field, where the coal seams were bigger and richer—one-fourth of Colorado is underlain with coal deposits—whose camps at Delagua, Aguilar, Forbes, and Ludlow were filled with an even greater variety of immigrants, the days of intense industrial warfare have been studied and chronicled by Upton Sinclair, George McGovern, Barron Beshoar, and others.

But it was to the tiny, relatively unknown Northern Coal Field that union men came first to organize the miners. It was in the northern field where Colorado's long coal strikes started and spread southward.

Because the original residents of the northern field came from such varied backgrounds, it is not surprising that distrust, even hatred, erupted into violence from time to time. When the Ku Klux Klan enjoyed a brief period of power in Colorado during the 1920s, burning crosses and silent parades of men in white sheets were not uncommon in Louisville, Lafayette, and Erie.

In Boulder, the county seat nearby, coal miners were not welcome. After all, the "dirty coal diggers" had strange ways and often spoke in foreign tongues. The fledgling college town feared that men with lunch pails would somehow downgrade and befoul the community.

In turn, the miners resented Boulder's outspoken attitude of superiority and contempt. That resentment toward Boulder is still apparent today and has its roots in those days when families, fresh from the old country or on the run from deteriorating conditions back East, came to Colorado for a fresh start.

Despite the hard times and the despair that came from knowing they were taken advantage of, many of these early residents survived and prospered. Perhaps such songs as this one, sung to the tune of the "Battle Cry of Freedom," helped along the way:

> The union forever! Hurrah, boys, hurrah!
> Down with the gunmen, up with the law;
> For we're coming, Colorado, we're coming all the way,
> Shouting the battle cry of union![6]

All is quiet in the Boulder valley.—*Etching, Western Historical Collections, University of Colorado at Boulder*

CHAPTER ONE

> It seems like providential arrangement that so many valuable mines of the precious metals, and such vast coal measures, should have been placed in proximity so close. At the Marshall and Black Diamond Banks, the first settlers found coal cropping abundantly to the surface, and soon learned that it was cheaper and better fuel than wood, however easily obtained.[1]

Before the fall of 1858, the Boulder Valley lay quiet, undisturbed, and undiscovered, except for the seasonal encampments of the Southern Arapaho Indians, who regarded the protected area as an ideal winter stop. Tribes of Ute Indians, who preferred mountain camps, sometimes ventured into the valley, if only to fight the Arapaho, their long-time enemy.

Game was plentiful. Buffalo and antelope fed on the valley's short, thick grass. Black bear and mountain lion were evident, a scene in considerable contrast to the one painted by explorer Major Stephen H. Long in 1820 as the inhospitable and unprofitable "Great American Desert," an internal frontier of little consequence.

A few mountain men and early travelers wandered about and camped in the area before the rush for gold. If the mountain men studied the coal outcroppings, even chipped off a few chunks for fuel, they didn't tell anyone about it. Seldom did these early wanderers commit their observations to paper, as they were "an uncommunicative lot and had no precise information about what they had seen."[2] Despite the silence of these early visitors,

> anyone familiar with the occurence of coal in most parts of this continent finds it impossible to believe that the earliest travellers through the forests did not see them plainly and that such keen observers as these men were, whose very lives depended daily on their powers of observation, missed the many evidences of coal seams along outcrops, in streams or under uprooted trees.[3]

Geologist Arthur Lake believed that Colorado coal was found by chance:

> Nearly all the coal that has been opened up in Colorado was first discovered by accident, by the digging of a well or a foundation, or by the outcrop of the coal itself, plainly seen along the banks of some stream.[4]

Mountain man Jim Baker.—*Photo, Colorado Historical Society*

Evidently the Indians did not regard the coal banks as a valuable resource but referred to coal as the "stove that will burn long forest fire."

Farther west, the fur trade had played out. Local forts up and down the South Platte River, built in the 1830s and 1840s to receive fur peltries, had closed and were crumbling.

Shortly after first settlement, however, the year-old *Rocky Mountain News* reported on August 13, 1859, that coal and limestone had been found some miles north of Denver City. A freshly chipped-out chunk of coal now graced the editor's desk. By late September, the *News* had more precise information for its readers on the discoveries, locating the coal seams eighteen miles north of Denver on the gently rolling plains. Claims to these coal lands were being filed upon, and the *News* reported that coal would be "offered in our market in a few days for one dollar per bushel."[5] (Twenty-five bushels are equal to one ton.)

Coal is mentioned in another of the *News's* reports to Denver readers:

> At the crossing of the Vazquez river, near the west foot of the table mountain, we found the thriving village of Golden City, scarse two months old, yet containing over

> thirty finished houses, many of them large and substantial, and twice as many more in the course of construction. Two steam-saw-mills were puffing away, ripping huge pine lots into building materials of all kinds. At a blacksmith's forge we noticed a pile of stone-coal of fine quality, obtained in the vicinity.[6]

One of the Northern Coal Field's first coal entrepreneurs was the aging Indian scout, Jim Baker, who had traveled in an earlier, more romantic time with Jim Bridger and the American Fur Company. Baker settled briefly in 1859 near what would become Erie, fifteen miles east of the mountains, on the west bank of Coal Creek. He had come west when he was fifteen years old in the 1830s, had participated in many a rendezvous, but now was an "old hoss" at forty-one. He was mourning the death of one of several Indian wives and was casting about, never in one place for long.

His Erie claims included a coal seam which he called Baker's Bank. The mountain man's new neighbors purchased the pieces of coal he had chipped away—it was not yet a proper mining operation—and, from time to time, he hitched his horses to a wagon and carted a load of coal into young Denver, where he sold the precious fuel at the going rate of twenty-five dollars per ton, a price that would soon rise to forty dollars per ton.

Baker also maintained a toll bridge, Baker's Crossing, spanning Clear Creek to the south. An Indian woman or two collected toll charges for him during his frequent absences. It was not long before the former adventurer tired of business life and sold his coal interests to move on. Too many people were on the scene, he said; he needed more "elbow room." Baker went to live with the Shoshone, his wife's people, and eventually ended up in Savery, Wyoming, in 1873.[7]

Those who developed wagon mines similar to Jim Baker's found the work hard and usually unprofitable. They were too early. "In locating a coal mine with a view to development," said Professor Lake in 1889, "the first thing to be considered is its proximity to a railroad, for without a railroad a coal mine is helpless.[8] The iron horse would not arrive in the northern field for a number of years.

Eastern capital was tied up in the Civil War, and the railroads had not stretched far enough west to transport the coal to market or to fuel their own engines. The would-be coal magnates continued to cart their brown-black chunks along the "oxen trail" to Denver, hoping for better times ahead.

Some despaired of waiting. George Gilson did not have the faith that "the money to open the mines must come, sooner or later."[9] He homesteaded land in 1859 near the present Erie and, like Baker, sold chunks of coal to the settlers. Discontented with Colorado's arid climate, Gilson decided he could not make a good profit from his claim. The railroad would never come.

One year later, he traded his acreage to Captain Ira Austin for a flock of sheep. Realizing that the sheep would slow him down, Gilson traded again, this time for horses and wagons. Grumbling about his ill fortune, he went back to the States to rejoin his family.

Eight years later, George Gilson came west again, this time with a caravan of prairie schooners. Curious to see what had happened to his former property, Gilson stopped off and was astounded, perhaps a bit chagrined, to discover a flourishing Boulder Valley Coal Company, which had been in operation for two years. The growing town of Erie had become a stage stop on the road to St. Vrain and had been christened by the Reverend van Valkenburg after his native Erie, Pennsylvania.

When Gilson heard local talk of a railroad that would soon chug into Erie, when he heard that Captain Austin planned to sell at least part of his holdings to eastern interests, the former coal merchant departed in some confusion, no doubt wondering what he had been thinking of in 1860.

At the southern end of the coal field, five young men had staked a parcel of land in the summer of 1859 and filed their claim with the Great Western Land Claim Association on July 20.[10] The new coal seam of the five prospectors—Henry C. Clark, Charles Wagen, John Cochran, C.C. Houchet, and one Barrett—sloped into the low hills and ran below a limestone ridge, an area that would in time become Marshall, a spot five miles southeast of the new settlement at Boulder City.

First they built a cabin as if they meant to stay. By October, they had begun to chip at the coal seam that tilted out of the ground nearby. On the thirteenth, the would-be coal merchants had a visitor, a traveler signing himself "L.N.T.," who wrote back to the States that his party had stopped with the young men:

> We accepted their hospitality and found them to be quite intelligent. Mr. H.C. Clark, one of the party, showed us the mine. . . . We found it to be very rich bituminous coal, free from sulphur, and strongly impregnated with rosin. There are three distinct leads, segregated by granite rock. The whole mass has been thrown from its natural horizontal position up an angle of thirty-five degrees by some convulsive upheaval. The price received in Boulder and Arapahoe [a camp three miles east of Golden, later a railroad stop] is seventy-five cents per bushel.[11]

What ever happened to Henry Clark and his fellow coal miners is lost to history. Soon thereafter, Joseph Marshall was in the area studying the lignite outcroppings as well as a red-brown earth he guessed probably contained iron ore. Perhaps, thought Marshall, this was the place to build a blast furnace to convert the reddish hematite into castable iron. As Marshall traveled back to his Larimer Street foundry in Denver, which he had purchased in December 1860, he knew that he would come back to the coal fields. He started filing claims.

Another early claim filer, William Kitchen, watched his fellow settlers pick up coal chunks to heat their newly built cabins—the virtually treeless area produced more coal than wood—and thought to himself, what happens when they can't pick up coal for free? He, too, felt he had a business opportunity. He made claims to the lands around an outcropping of coal, near Marshall's future blast furnace site, and named his find the Washington lode.

Kitchen's neighbors, however, continued to call it Kitchen's Bank and bought fuel from him, as he predicted, when they could no longer find loose pieces of coal of their own. (There weren't many chunks loose on the ground, as the coal turned to powder rather quickly when it was exposed to the air.)

The new coal man hitched up his team and wagon, as Henry Clark and Jim Baker had done, loaded with newly dug coal, and took off for Denver. Sometimes he hired Boulder City resident Marinus G. Smith to cart the coal to market in his Marinus Express wagon. Smith regularly journeyed to Denver to get young Boulder's mail. Local families began to contract with Smith for coal delivery to their homes.

Marshall drawing, Denver Public Library

But Kitchen, like George Gilson, found working a wagon mine both tedious and unprofitable. After a few years, he sold his holdings to Joseph Marshall, who did not stay in Denver long. While he was there, however, Marshall built his first blast furnace near the Larimer Street foundry, concentrating ores dug on his new claim near South Boulder Creek. His first production of pig iron was delivered by wagon freight to Central City and other mountain camps above.

At some point, Marshall and his partners cast a two-inch bore cannon, manufactured from the hematite ore. The cannon was first fired in 1861 to celebrate Independence Day.

In 1862, Marshall sold his portion of the Denver foundry to his partners Augustine Gallet Langford*, William L. Lee, and Mylo Lee, who promptly moved the business to the mountain village of Black Hawk. All of the principals were frustrated in their attempts to transport goods back and forth from the lowlands to their mountain operation. In no way could the trails to the gold mines above be described as proper wagon roads at this time.

By 1863, Marshall was busy building a new cold blast furnace, the Belle Monte, on his South Boulder Creek lands. It was not long before a New York financial firm issued a business prospectus describing Marshall's operations, calling the settlement Belmont. Soon that name appeared on maps of the period.

The Belle Monte, of course, required water for power. At first Marshall bought water from his neighbors, the McGinn brothers, who held early water rights and had built a ditch off the South Boulder in 1859. Eventually, Marshall built his own ditch from the creek to the furnace, where he constructed a huge waterwheel to power his project.

When the McGinns learned that Marshall would soon need some thirteen to

---

* Joseph Marshall had known the Langford family in St. Paul, Minnesota, in 1857. Their grandfather, George Langford, had built a blast furnace in Westmoreland, New York, near Utica, as early as 1801. This furnace he called the Hecla, a name which resurfaced in the 1870s for several Colorado coal and hard rock mines, most notably Marshall's blast furnace along the South Boulder and a future coal mine near Lafayette.

Joseph Marshall experimented with many ores and machines.
—Photo, Colorado Historical Society

fifteen bushels of charcoal to reduce each ton of iron ore, they went into the wood-gathering business. The brothers had to go as far as the mouth of Eldorado Springs Canyon, two miles west, to find sufficient wood to burn in the furnace.

Some of Marshall's neighbors were flooded out in 1864. Alexander Douty and his son Sylvester had built a grist mill along the South Boulder in 1861 and enjoyed a brisk business until they were washed out. Rather than rebuild on the same site, the Doutys moved to the mouth of Boulder Canyon for another try.

By 1864, Marshall had produced his first pig iron with the Belle Monte. Even though the *Rocky Mountain News* of July 7 pronounced the Belle Monte Iron and Coal Company a "success and a valuable institution to our territory," the furnace blew out and collapsed. Marshall did not despair and rebuilt the furnace with better hearths. The new structure was twenty-two feet high and seven feet in diameter. Eventually, the foundryman produced some two hundred tons of acceptable pig iron at a cost of $75 per ton, which he turned around and sold in Black Hawk for $160 to $200 per ton.

No wonder Joseph Marshall has been described in subsequent articles on his work as the "Tubal Cain of Colorado."[12] (In the Bible, Tubal Cain was recorded as the first worker in brass and iron.)

Marshall's dream was to bring large quantities of gold and silver down from the mountains on the new Bear Canyon road, which was built south of Boulder. He planned to reduce the precious metals in the Belle Monte. Unfortunately, the Bear Canyon road washed out almost as soon as it was constructed; undaunted, the road-builders started again and again, only to be washed out over and over.

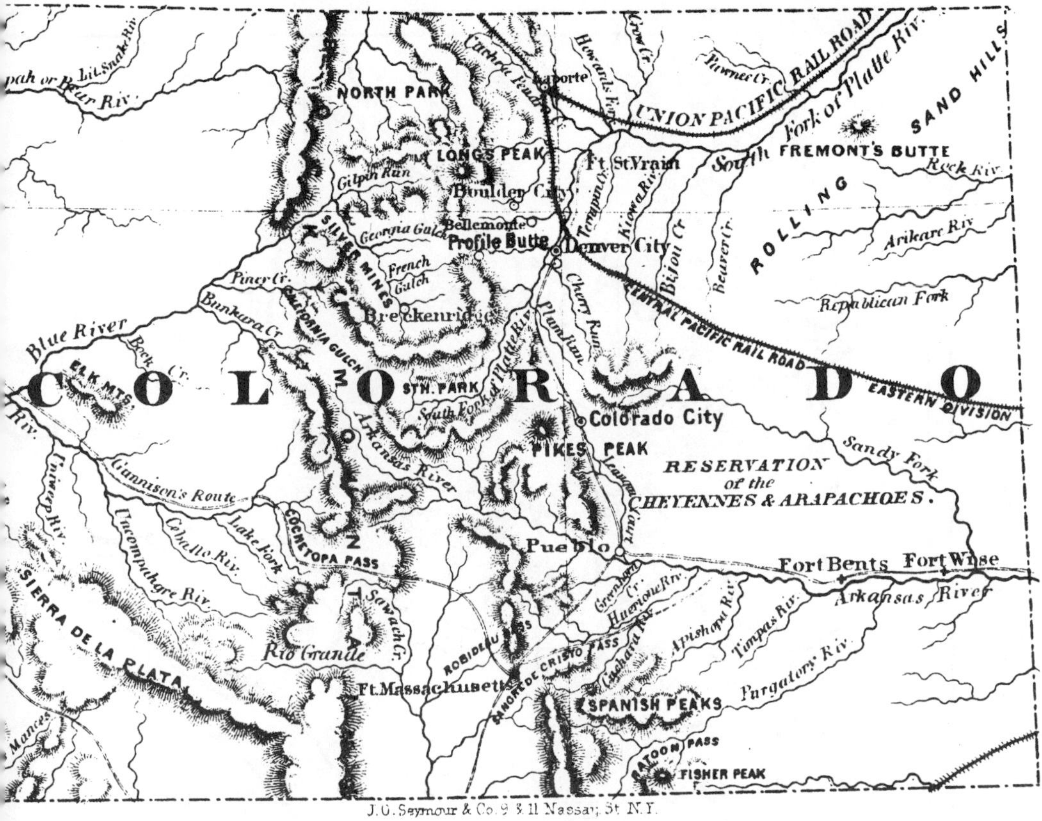

Marshall's Bellemonte is on this early map.—*Map, Denver Public Library*

When traveler-writer Bayard Taylor visited Colorado in 1866, he wrote back to the States that he was "fast inclining toward the opinion that there is no American Desert on this side of the Rocky Mountains."[13] The area did not seem as drear as previously reported.

Taylor found Marshall's operation as he traveled through the area:

> This is a charming little valley, at the base of the mountains. The outcropping of limestone, and the black piles of the mouths of coal drifts indicated our approach to it. On dropping into a little winding hollow, we soon saw the massive smelting furnace surrounded by clustered cabins.[14]

At the time of Taylor's visit, Marshall, in addition to his furnace operations, had opened coal seams two years before on the land he bought from Kitchen to start a proper mining enterprise, the first in Colorado.

Taylor continued to report to his Eastern readers:

> Mr. Marshall, the proprietor, received us at the door of his residence, and after dinner, piloted us to the furnace and to the mines. There are eleven veins of coal, varying from four to twelve feet in thickness, in the space of half a mile; iron ore or a richness of fifty percent, just beyond it . . . and the best limestone, in almost

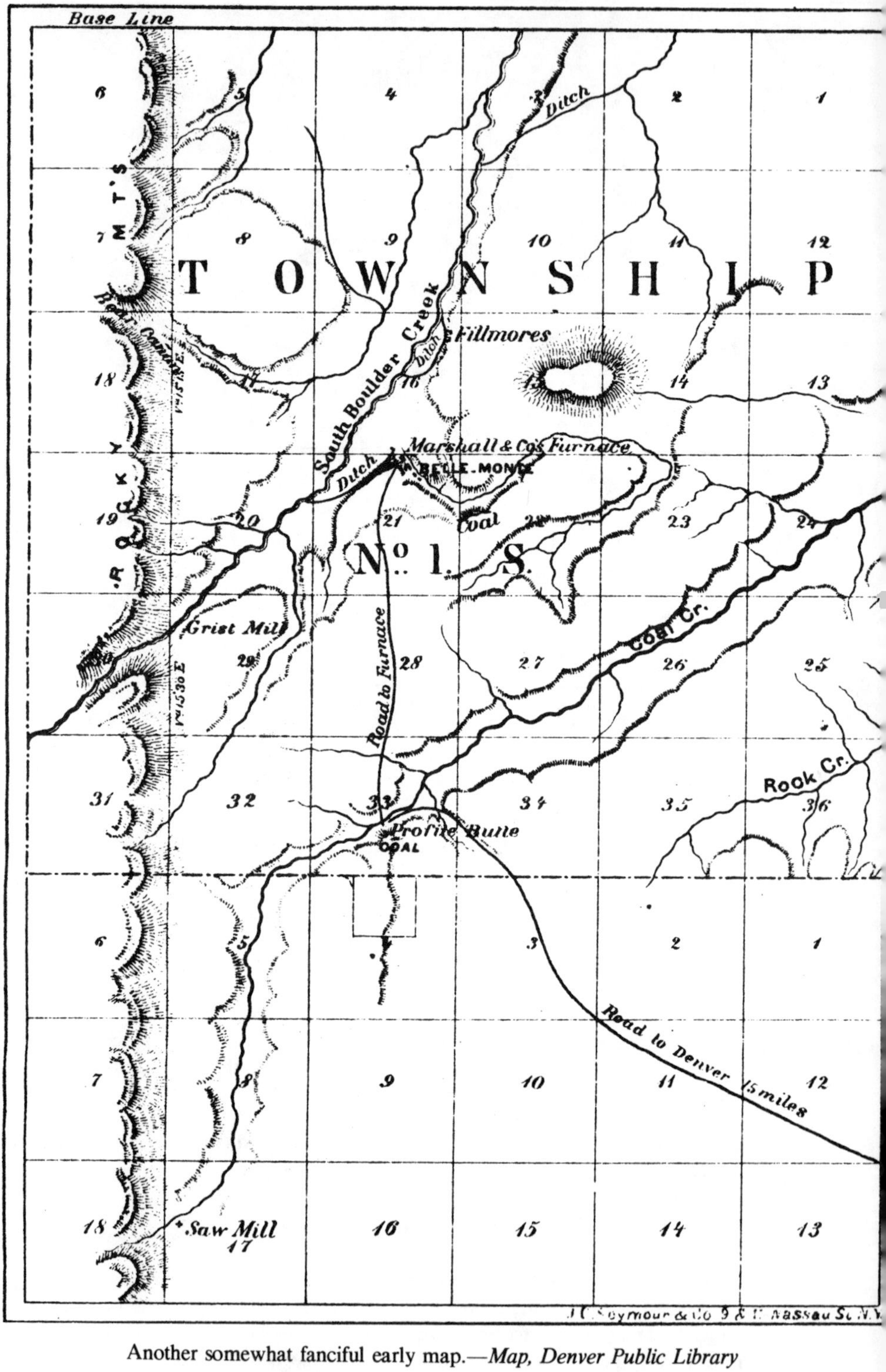

Another somewhat fanciful early map.—*Map, Denver Public Library*

inexhaustible quantities. Mr. Marshall, however, has only experimented with the native ores sufficiently to establish their value. He finds it more profitable to buy up abandoned machinery at a trifling cost, and recast it. The furnace is not only substantially but handsomely built, and has thus far done a thriving and successful business for its owner.[15]

Perhaps Taylor was too fulsome in his praises and overly optimistic about the Belle Monte. Shortly thereafter, Marshall abandoned the pig iron project and devoted himself exclusively to the establishment of the Marshall Coal Company and the development of the Black Diamond and Marshall mines.

In addition to Kitchen's Bank, he had increased his coal field holdings to 1,480 acres, thanks to an 1868 grant signed by President Andrew Johnson. As his operations grew, more and more families settled nearby. No longer called Belmont, the camp at Marshall, although never incorporated, soon grew to more than six hundred residents who supported a school and three saloons.

Other pioneers now settled the area with no immediate thought to sink a coal shaft. They were more interested in growing wheat and developing dairy herds on the land Stephen Long has dismissed as forever unprofitable. Henry Adolph Waneka came from Germany in 1860 to farm; since he had not brought his family as yet, he "batched" it the first winter in a cave he found, located southeast of Louisville.

That same year, William Charles Hake and his family came by covered wagon from Superior, Wisconsin, to homestead in what they called the Coal Creek Valley. The area would be called Superior by the late 1890s. The Hakes planted apple trees, planned their crops, and started raising stock. Despite the fact that the floodwaters of 1864 had uncovered a massive coal seam on his land, farmer Hake apparently did not realize that he had settled on top of what would one day become the prosperous Industrial mine. It was not until the 1890s that Hake did anything about it.

In the late summer of 1863, newlyweds deLafayette* and Mary Elizabeth Foote Miller came west along the Smoky Hill route with a fifty-wagon ox train from Missouri. The young couple brought three wagons and twelve oxen and put down first roots on a 160-acre homestead in Burlington, near the future Longmont, a few miles north of the coal fields. At some point, the couple imported the area's first threshing machine, freighted in pieces from the States.

One year later, the Millers realized that they had been "taken" in a Burlington land scam, so they moved south to purchase Rock Creek House, an inn and stage stop north of Rock Creek, along the Overland Trail, earlier called the Cherokee Trail, between Denver and Cheyenne. The Pony Express stopped at Rock Creek House. Those who were bringing Texas cattle to market stopped there for a rest. Mary Miller remembered the overnight visits of General U.S. Grant and General W.T. Sherman. Mary's sister and her husband bought and operated the next stage stop south, Churches, near the present Broomfield.

Seven years later, the Millers decided that the homestead lands near what would become the town of Lafayette looked promising for farming, despite the lack of trees in the area. They bought 320 acres. They had no thought at that time of

* No one called him deLafayette, just plain Lafayette.

Mary Elizabeth Foote and Lafayette Miller were married in Buchanan County, Iowa, December 24, 1862.—*Photo, Lafayette Public Library*

An early shot of Rock Creek Station, which Mary and Lafayette Miller operated for a time.—*Photo, Lafayette Public Library*

opening a coal mine on their farmlands.

Instead, Lafayette operated butcher shops in Erie and in Boulder with his brother-in-law. Mary reared their growing family of five boys and one girl. They left the farm in 1874 and moved into Boulder, where Lafayette interested himself in politics. He was elected town trustee and was narrowly defeated for mayor. Mary Miller educated herself and her children at home by reading aloud an entire set of encyclopedias, little realizing that, in a few years, she would be widowed, she would found a town, and she would become a prosperous banker. Meanwhile, the Millers increased their farm holdings to some 1,280 acres and called the acreage Willow Glen.

Another early resident, David Kerr, Lafayette Miller's uncle, arrived in 1864 with his family to settle near what would become Louisville. Kerr bought farmland five years later under a grant signed by President U.S. Grant and planted his first wheat crop, unaware that underneath his soon-golden fields lay valuable seams of coal.

Those who had come to stay were in place. The go-backers and adventurers had left for the States or the Black Hills or Montana Territory. Samuel Bowles wrote in 1869:

> The vicious and the loafers, the gamblers and the murderers, have mostly "moved on;" what is left is chiefly golden material; and the mines and farms of Colorado, all alike and together, are in healthy and promising condition, and insure for her a large growth and generous future. The two things she lacketh chiefly now are appreciation at the East and women; what she has of both are excellent, but in short supply; but the Railroad will speedily fill the vacuum.[16]

Those who had chipped away at the drifts of visible coal outcroppings to fill their wagons with the black fuel for market had either sold their claims to those with more experience in coal mining or connected with men who knew how to develop a proper mine.

By the end of the first decade of settlement, the Denver and Pacific Railroad had completed its line from Denver to Cheyenne, Wyoming. Within a year or so, a fifteen-mile spur would be built west from Hughes Station near Brighton to the developing coal fields near Erie. The railroad needed the coal for its own operations as well as to supply markets in Denver and as far east as Kansas City.

Those local residents who owned the Boulder Valley Coal Company, now on solid financial ground, were the first to sell half of their interests in 1868 to eastern businessmen and half to the Kansas Pacific Railroad Company. C.F. Wallace became both the Boulder Valley mine superintendent and the station master for the railroad. This double role was not unusual and was repeated over and over again as eastern railroad interests bought more and more land in the Northern Coal Field.

Thus, fewer and fewer coal mines in Colorado were owned by local residents. Absentee ownership would soon become an important factor in the deteriorating relationship between mine owner and worker. Only a few wagon mines operated now; most lay at the northern end of the field.

Knowledgeable coal operators looked for the wider coal seams that lay beneath the ground. Double entries, or tunnels sloping into the low hills of the area, became a common sight; they were easier to develop than mines that required a vertical shaft dropping several hundred feet beneath the ground. Such operations waited for the expertise of men who had sunk shafts in the East or in northern Europe.

More and more experienced coal miners from England, Wales, Scotland, and Germany came to the northern field now; in addition to their special knowledge of coal, they also carried in their heads blueprints for a miners' union, and they didn't mind talking to others about their dreams. Within a few years, however, it would become unwise to talk freely about the organization of labor.

Back in the States, a group formed in 1869 that would affect both miner and coal operator in the Northern Coal Field. Within a few years, organizers from the Knights of Labor would be on the scene in Erie and in Lafayette, the first union men to arrive in Colorado. By 1886, one million men in the United States would become members.

Inside the Vulcan mine.—*Photo, Lafayette Public Library*

CHAPTER TWO

> Coal has been the chief inducement for the early construction of railroads into the County.[1]

The 1870s opened with considerable optimism in the Northern Coal Field. The Civil War was over, thus freeing eastern capital for investment in the West. The "iron horse" was on its way. With completion of the Denver Pacific's Boulder Valley Railroad to Erie in January 1871, prospects seemed unlimited for the new mining industry. Spring wagons, buggies, hacks, and carriages from all over the Boulder Valley met the trains at Erie. Tickets from Erie to Denver cost $2.25 per passenger. Railcars filled with coal bound for the Denver and Cheyenne markets rumbled out of the little town daily. All was bustle and enthusiasm. Before the spur to Erie was completed, Denver citizens were able to buy only 15,860 tons of coal each year to heat their homes. In 1871, with the rails stretching into the northern field, Denver alone purchased 68,540 tons of coal.

Geologist Ferdinand V. Hayden, not known for modest predictions, proclaimed to the papers:

> Nowhere in the world is there such a vast development of the recent coal measures, and in few places is their existence more necessary to the advancement and improvement of the region in which they occur.[2]

These coal deposits were mentioned as an asset to those who were forming colonies to the north and east of the field. The letters and brochures describing these new settlements—Greeley's Union Colony, established in 1869, the Chicago-Colorado Colony at Longmont, formed two years later, the St. Louis Western Company at Evans, also settled in 1871, the Platte River Land Company near Platteville, another 1871 establishment—commented on the proximity of fuel that would be at the disposal of the new residents, an exaggerated claim at best. Geologist Hayden allowed that the new colonists would have to wait for new railroad lines into the colonies to be assured of coal supplies:

> It was supposed, and the report circulated at the time we came here, that the whole country around was underlaid with coal. A company was formed in the fall of 1870 to prospect for coal, but made no discoveries of any near that was good for anything. Professor Hayden soon told us that the Poudre Valley was eroded below the lignite coal deposit. . . .No depth of digging would find coal under us as we found to be true in digging our artesian wells, one of which was put down 2,700 feet and no coal met.[3]

Visitors to Boulder County's second fair in 1870 gawked at a mammoth piece of coal from Joseph Marshall's mines; weighing some eleven thousand pounds, the chunk measured six feet across and four feet, eight inches deep. Another specimen from the southern end of the field near Ralston, a two-ton piece of coal from Murphy's Bank, was also on display.

Very quickly, Colorado's coal industry developed sophisticated methods to get the coal out of the ground. The relative softness of northern Colorado coal led to the use of the "room and pillar" method of underground development. Two parallel tunnels, called entries, fanned out, each set at right angles from the bottom of the shaft. If the tunnels were long, cross-cuts were made to connect the entries to help the flow of good air down the passageways, although "good air" was seldom in sufficient supply. Near the bottom of the shaft, at least in the better-developed mines, was a huge fan that sucked good air from the top of the shaft into the various tunnels. When fully developed, the "city" underground resembled a human anthill.[4]

From each entry, "rooms" were cut out, some ten to thirty feet apart, providing a "pillar" of coal or rock between each chamber. Usually two miners worked a room; they regarded that area as their special province until they removed all the coal at that spot; as they retreated from each "finished" room, they usually pulled down the pillars behind them.

"Shooting off the solid," a method of blasting the coal out with explosives, could not be effectively used in northern field mines as it was in the East. Since northern field coal was not as hard, the blast would tend to pulverize the mineral and make it worthless for market. Furthermore, this method, fast and efficient to be sure, fractured the roofs of the rooms, causing cave-ins.

The northern field miner would approach the vein of coal with his pick and, lying on his side, most often in cold water and debris, would hack out an undercut, called a "kerf," beneath the vein.

Then, pulling himself out from the wet undercut, he stood to drill holes in the seam with his hand-held breast auger. The holes completed, he placed a tube of black powder into each hole; into the powder he slipped a long copper needle—copper did not produce dangerous sparks—and dummied it into place with mud and coal dust. When the copper needle was removed, a squib, designed "like a pop bottle rocket,"[5] was put in place, which, when lit, "flew down the passage to explode the powder."[6] Down came the coal in loadable chunks. Years later, dynamite was used, although it had to be placed in such a way that it would not disintegrate the soft coal.

Eventually, one man specialized in this kind of work; the shot firer would move from room to room, setting off squibs to discharge the coal. When the mule driver brought an empty car—later, it was the motorman—the miner would carefully stack the car so that it would carry its largest possible load. Larger chunks at the bottom, then filled in with smaller pieces.

With the first tunnels and shafts underground came the certain realization that the nature of the sub-bituminous coal and neighboring lignites was such that the coal could not be stored for future delivery to customers. Underground, the mineral had a fairly high moisture content—15 to 20 percent. Above ground, however, the fuel became unstable, soft, and friable; it soon crumbled to slack and

Marshall boy stands near coal mines burning underground.—*Photo, Western Historical Collections, University of Colorado at Boulder*

powder, unlike the harder anthracite coal mined elsewhere. Moreover, coal from the northern field was highly volatile and combustible, a condition that prevented stockpiling for later delivery as was the custom in coal yards back in the States.

Below ground, the black mineral was equally dangerous; its dust collected in piles, making it difficult to work, even breathe, in the chambers. Coal burns just as efficiently underground as it does in a stove or locomotive. In 1870, the first known underground fire started to burn in Joseph Marshall's Peerless mine. (When Colorado's first inspector of coal mines, John McNeil, made his 1884 report to Governor James Grant, the inspector mentioned that the Peerless was still on fire after fourteen years. In the 1980s, the Peerless and other nearby chambers of the Marshall mines still smoldered beneath the ground, flaming above the land's surface from time to time when fissures supplied enough oxygen to fan the fires.)

Because northern field coal could not be stored, the operators consistently "miscalculated" the tonnage needed for their markets and overproduced. This miscalculation became a wasteful habit, one that would cause trouble between labor and management throughout the state, even after the turn of the century.

Thus, coal mining was a seasonal occupation at best; summers found the miners idle, tending small gardens for food to feed their families. Some men left the state to work as gandy dancers* on Utah or Wyoming railroads for the season. When the first snowflakes signaled cold weather, they were back to work in the mines.

Typically, when the operator had too much coal on his hands, he would lay off his crews. If the men objected, the superintendent threatened to hire a new crew. Labor was not in short supply as more families settled the area.

In February 1871, officers of the Boulder Valley Coal Company decided that the miners at their Briggs slope mine near Erie should screen and sort the coal for

---

* Men who worked on a section crew for the railroad. Perhaps the term "gandy dancer" was derived from the Gandy Manufacturing Company, Chicago toolmakers.

market for the same daily wage—$1.25. The men wanted an additional fifty cents per day for screening. The management said no. That decision was all that was needed to start the first coal strike in Colorado.

Leaders Thomas Higgins, Mike Kirby, and L.N. Herring led defiant strikers on a march to Erie on February 25. Along the way, fistfights broke out between striking and nonstriking miners. Sheriffs from two county jurisdictions arrived. Weld County Sheriff George Brush formed a posse to break up the rioting men. The three leaders plus nine other miners were arrested, loaded into a boxcar, and taken to the county jail at Evans, near Greeley. The strikers were tried, found guilty of conspiracy, and fined. After one week, the Briggs mine reopened and the now "docile" men went back to work, forced to screen coal for the same wage.

The times were not right yet to form a proper union, but a number of the men continued to talk about the possibility, very quietly. Any man whose interest in labor unions came to the attention of the coal operator was dismissed and replaced with a more "reasonable" worker.

New mines opened, more miners arrived to settle the area, and, by the end of 1871, the railroad car had pretty well replaced the team and wagon as a coal carrier. Despite the first labor troubles, wages increased from $1.25 per day to $1.25 per ton, or about $3 per day.

A feeling of confidence spread from management to the workers and lessened the immediate need for unionization. Good times were coming. The sight of busy steam engines hauling coal to market heightened the rosy glow of financial boom.

In 1872, a county newspaper reported to its readers that "Erie is becoming one of the most lively little towns in this part of the country."[7] A year before, a mere fifty persons populated the immediate area; the community had been laid out on the west bank of Coal Creek, although it would not be incorporated until 1894. The main street was now called Briggs; other lanes were named Moffat and Cheesman to honor officers of the Boulder Valley Coal Company.

The paper also reported that "a good many buildings are being erected, and men with families are moving in. A flourishing school is supported, numbering thirty-three pupils."[8] The new townspeople had planted saplings along Briggs Street. The question of fresh water, or any water supply for that matter, was temporarily solved by the sinking of artesian wells. The Union Pacific had also struck a well for its water and allowed citizens to use that supply. Many housewives used the waste water pumped out of the mines for the weekly laundry.

Soon a cheese factory was established between Erie and Valmont. Several hotels were constructed to house railroad men and bachelor coal miners—the Central Hotel, the Agnew Hotel, and Erie House. The fourteen-room Erie House was owned for a time by the Reverend van Valkenburg, who apparently mixed religion and finance with no ill effect. During his term as Erie's second mayor, the minister also helped to form a local lodge of the International Order of Oddfellows. Other fraternal orders—the Macabees, the Rebekah Lodge No. 29, Rose of Sharon—an important part of coal town society, were established in the next few years. Erie farmers joined Grange No. 35.

Since the little town was on the railroad line, residents built the Erie Opera House, located near the lumber yard, to take advantage of the presence of traveling light opera companies and other circuit entertainments. The Erie Brass Band,

The new town of Erie, June 1876.—*Photo, Denver Public Library*

The Angove-Hershey Feed Stable.—*Photo, W.A. Warnock*

Early resident William Angove poses with daughter Josie and wife Elizabeth in front of new home.—*Photo, W.A. Warnock*

Members of the famous Erie Brass Band proudly pose.—*Photo, Daisy Bracegirdle Swallow*

established in 1875, gained a reputation for spirited concerts and parades. There were dances, foot races, trap shoots, and baseball games. A dirt track was built to accommodate another popular sport of the day—horse racing. Just before Christmas, one decorated tree was placed at the town's center—individual households did not have a tree—and everyone got ready for a holiday ball.

Small businesses proliferated—meat markets, hardware establishments, grain elevators, saloons, and a boot shop operated by M.S. Cummings and Andrew Hamron with the imposing name The Knights of the Last, whose products were designed "to last until lasting a last becomes almost an endless last and lasting on the thanks of those who patronize these Knights of the Last."[9]

Mrs. Ella Leyner, whose "skill and grace"[10] was renowned, set up shop as a milliner. By 1900, the Kuner Empson pickle factory was in business at the south end of Erie.

At the southern end of the coal field, Joseph Marshall could not wait for the railroad to be built to his mines. In 1872, he ordered a steam engine from England, a twelve-ton wagon designed to pull his coal carts along the new road to Golden, which was being constructed by John W. Nesmith, the inventor of the spark resister for locomotives.

When the steam engine arrived, the Marshalls hosted a grand party to celebrate its delivery. At first, the engine carried from ten to twelve tons of freshly cut timbers for Nesmith's bridges under construction on the Golden road. When the "highway" was finished, the British wagon carried coal to Golden and returned to Marshall—a "speedy" trip of some twenty hours.

From the road, Marshall looked up at the mountains, still dreaming of freight loads of gold and silver ore moving down the Bear Canyon road from Black Hawk. Like the blast furnace, however, the steam wagon proved to be one of Marshall's less successful ventures. He was forced to wait for the completion of the

Marshall is becoming a community, 1875.—*Photo, Western Historical Collections, University of Colorado at Boulder*

Folks pose in front of the Central Marshall Saloon. Note the wooden kegs of beer out front.—*Photo, Boulder Historical Society, Carnegie Branch Library for Local History*

The meat market at Marshall (Langford).—*J.B. Sturtevant photo, Boulder Historical Society, Carnegie Branch Library for Local History*

Golden, Boulder and Caribou Railroad to his coal measures before he could successfully expand his operations.

Another coal merchant, the patrician William A. Davidson, one of Boulder Valley's few settlers who had a preparatory school background—he went to Phillips-Andover—founded his coal business in 1873. The Davidson Coal and Iron Company was incorporated with Colorado Central Railroad officers W.A.H. Loveland, C.C. Welch, and H.M. Teller as principals. Earlier, Davidson had dabbled in gold mining ventures, operated Boulder City's first general store with partner Samuel Breath, and farmed a six-hundred-acre spread, four miles east of Valmont.

Davidson produced one of the area's first strawberry crops. Later, he planted extensive acres of sugar beets. He found coal on the company's eight thousand acres after striking a shaft thirty feet below the ground. For their water needs, the mining group built a thirteen-thousand-dollar canal from South Boulder Creek to the coal fields, eleven miles away, a grand project for its time.

One year later, Davidson platted a town in his name, northeast of Marshall, which immediately "boomed" to twelve buildings and included a blacksmith, post office, saloon, and railroad depot for the Colorado Central Railroad. Several years earlier, officials of the Colorado Central, in order to compete with the Denver Pacific, as well as the Union Pacific, had worked hard and speedily to complete a line to Golden. With the expanding coal sites on their minds, they extended the road by April 17, 1873, from Arapahoe Junction, three miles east of Golden, to snake its way through the coal fields to Boulder, then straight on to Longmont.

Traveling through the area on horseback that year, British gentlewoman Isabella Bird took note of a promising discovery of a coal seam in the valley but saw Boulder City as nothing but a "hideous collection of frame houses on the burning plain,"[11] and immediately departed for her beloved Estes Park.

When the Denver and Boulder Valley Railroad extended its line to Boulder City in 1873, more coal mines opened along its path. By this time, the railroad (a subsidiary of the Denver Pacific) owned more than eight thousand acres of coal land.

Near the Tabor Station stop, twelve miles east of Boulder, promoter Isaac Canfield and his brothers struck a coal seam they called the Rob Roy. Having little in the way of financing, Canfield wrote excitedly to one of his relatives who had just settled in the Union Colony at Greeley. He told the newcomer, H.T. West, that the prospects for coal in the Denver market were so promising that the Canfields would let West into the family business as a shareholder. All West needed to do was furnish $12,500 as well as an additional $18,000 in credit guarantees. The well-heeled West rushed down to inspect the Rob Roy. The family decided that the new stockholder should move to Denver and act as agent for Canfield coal. The town of Canfield was going to boom for sure. Even though it was not yet incorporated, twenty buildings were "almost" up.

But Isaac Canfield and his relatives had not reckoned with the possibility that a strike, which led to a series of strikes—sometimes referred to as the "Erie Wars"—might threaten their financial position.

Coal continued to be overproduced to such an extent that Superintendent W.J. Curtice of the Golden mine, a new shaft near Canfield's operation, ordered that wages for his one hundred miners be lowered to $1 per ton. Those who had been quietly talking about the need for a union came out in the open and charged into action. After a brief period of organization, miner George Ardat was elected head of a new local union on November 9, 1874. Curtice was intimidated and backed down.

But when another one hundred men at the Baker mine nearby expressed their desire to join the new union and also struck for higher wages, Edward B. Light of the Erie Coal Company responded by throwing the miners out and hiring a brand new crew. (The railroads had been so successful in encouraging immigrants to come to Colorado, even systematically recruiting them in Europe, that labor was in constant fresh supply.)

Threatened with violence between strikers and scabs, Light wrote to Territorial Governor Edward McCook asking for his help. He also placed passionate letters to the "People of Colorado" in the papers. All to no avail; Light was forced to negotiate with the strikers the following month, and, by Christmas, the original crew was back at work. The bewildered immigrants who had been brought in to break the strike moved on to what employment opportunities they could find.

This moderately successful unionization effort was the first of its kind in the Colorado coal industry, and it established a precedent for future organizing efforts in the Boulder and Weld County mines.[12]

Later, when union organizers moved into the Southern Coal Field, they used the same blueprints developed in the northern field until just after the turn of the century.

Miners in the Lafayette area watched Erie's labor success and decided that they too would form a local union. The new groups studied the emergence of the Miners National Association of the United States of America, hoping for eventual

The snow melts near a Marshall cabin.—*Photo, A.A. Paddock Collection, Boulder Historical Society, Carnegie Branch Library for Local History*

affiliation with a national union. Their hopes dissolved when the national labor group fell apart in 1876.

A few of the striking men, however, decided that continuing fights with mine managers were not to their taste. Eighteen miners elected to work for themselves and formed the Cooperative Mining Company in December 1875, having obtained mineral rights to the Mitchell mine, which lay just north of the Lafayette coal fields.

On March 18, 1875, Joseph E. Bates, superintendent of the Boulder Valley mine east of Erie, tried for a reduction of wages, saying that if the men didn't like it, he would follow Edward Light's example and hire a new crew. Bates was forced to capitulate, however, when he realized he also did not have public opinion behind his proposed wage cut. He reluctantly agreed to negotiate with the miners, and the men went back to work. But that was not the end of troubles for Joseph Bates.

By April 1877, officers of the national Knights of Labor felt the time was right for union expansion into Colorado; they sent agents Charles Vaughn and William Thomas to the Boulder Valley mine to "spread the gospel." Since most of the men already held strong views on the rights of labor, Vaughn and Thomas found an easy job of it, much to Joseph Bates's dismay.

The Knights of Labor were welcomed by the men even though the group's track record for effective organization wasn't good back east. The group's anti-business stance bothered some; the door of membership was open to all but not held as wide open for non-northern European workers. But everyone, it seemed, loved the passwords and secret ritual of the organization. They loved to sing the union's song:

> Storm the fort, ye Knights of Labor
> Battle for your cause,
> Equal rights for every neighbor
> Down with Tyrant laws.

The Erie Wars continued. On Friday, April 13, 1877, someone set off eleven kegs of explosives in front of the Boulder Valley mine, blowing up a fifty-foot area. On April 17, an enraged Superintendent Bates "helped" the labor movement along by announcing another drop in wages from one dollar to seventy-five cents per ton. He then ordered that organizers Vaughn and Thomas be thrown off the Boulder Valley mine property.

Two nights later, disgruntled miners came in to Erie from all over the northern field to take part in a torchlight parade. However, Judge Arthur N. Elliott of the Weld County court gave the county sheriff the power to arrest Vaughn and Thomas before the parade could form. Along with the sheriff came mine owners Sam Logan and Max Greenfield, General H.B. Bearce, and, of course, Joseph Bates.

Needless to say, this group was not well received by the angry miners, fifty of whom started toward the Baker mine with lighted torches in their hands. At the Baker, they set fire to a storage building and the winch house, but the enraged miners were sufficiently startled by the sight of the blaze that they dispersed without doing further damage.[13]

A few days later, a blast of unknown origin seriously damaged one of the tunnels at the Boulder Valley mine. (Some whispered that it was the work of the Molly Maguires, a secret Irish immigrant group of miners who had been using violent techniques, i.e., bombing, murder, intimidation, back east in Pennsylvania to combat oppressive working conditions.)

A desperate Bates hired twenty-five guards, commanded by General Bearce, who established themselves on the low bluff overlooking the mine, their Winchesters and Creedmoors at the ready. Bearce supervised the construction of a pole-built fort on the bluff and ordered his men to watch for troublemakers, much to the uneasiness of Erie residents nearby. A Boulder newspaper reported, "The boys in the fort practice target shooting a great deal, meaning only amusement, but it is very annoying to the inhabitants of the town, since the whole community is in range of the guns."[14]

Tension was only heightened when the guards stopped to interrogate Erie farmers who were bringing in their livestock from the fields beyond the mine.

Organizers Vaughn and Thomas were arrested, convicted of inciting a riot and threatening private property, and fined one thousand dollars. Since the Boulder Valley mine continued to pay the lower seventy-five cents per ton, a number of the miners sought jobs with higher wages elsewhere in the area.

Although it appeared that the union movement in Erie was dead, the Erie Wars continued, and subsequent actions of local courts, sheriffs, and editorial opinions of most local newspapers only served to coalesce the anger and frustration of the miners. It is possible that the extension of the Erie Wars led to the passage of a local ordinance that made it illegal to conceal about one's person either a pistol, a dagger, or a bowie knife.

Isaac Canfield was broke. He blamed the strikes when his company faltered. His Denver relative, agent West, ran out of credit, became destitute and confused, and entered a mental hospital. Canfield left the coal business and departed the area some time later; he would return in twenty-five years to participate in another Boulder Valley "boom" in 1901 as an oil-well driller.

By 1875, others had moved in to try their luck with Canfield's leavings. D.S. Woods became manager of the Rob Roy, which he renamed the Jackson, since it was now owned by the Jackson Coal Company. Former Wisconsin farm hand and sometime poet William O. Wise took over as superintendent of the nearby Star mine. (He was part owner with M. Brennan.) Wise and his brother, Joseph O.V. Wise, proprietor of the Standard Coal Company, B.M. Williams, and Charles F. Wallis platted the town of Canfield on December 27, 1875. They put the word out that Canfield was "destined" to become one of the richest coal mining regions in Colorado.

For a few years, Canfield did grow; a population of 398 supported a boarding-house, a meat market, and the Hilton John Saloon. The father of the Wise family, Oliver Ellsworth Wise, opened a blacksmith shop; a Wise Brothers General Store flourished for a time. As more mines opened nearby, the town gained a factory and the Butler Hotel. Yet another Wise family member, grandson William T. Wise, opened a feed mill.

In the 1880s, the community was able to come up with nine hundred dollars to build a school on land donated by J.O.V. Wise. William O. Wise successfully

Muddy street in Erie.—*Photo, Erie High School*

A miner's Erie home.—*Photo, Daisy Bracegirdle Swallow*

Erie's first hose team.—*Photo, Erie High School*

petitioned the federal government for a post office; he became its first postmaster in 1882. Two years later, he took time off from writing poetry to found the *Canfield-Erie Independent.* He also served in the Colorado General Assembly.

By the end of the 1880s, however, the only sign of growth in Canfield was the opening of a second saloon. The population leveled off, and the dream of Canfield as a major metropolitan center turned sour.

Not so with Erie. The little town was bustling. In 1877, the *Boulder County News* reported:

> Erie has grown up within the last few years. It now numbers six hundred inhabitants and the annual earnings of the coal miners amount to $120,000. The school has over one hundred pupils; most of them are children of the miners.[15]

By 1881, Erie was able to afford a new school for its children. The *News* also told its readers:

> The character of the people is excellent. It is as orderly, temperate, peaceable and good a class as is to be found in any town in Colorado. The Good Templars lodge numbers nearly a hundred, making the proportion of earnest temperance men larger here than in any other town in the state.[16]

An early holiday celebrated by Erie residents was Biscuit Day, a festival enjoyed some time around Labor Day. Baker Chris Miller offered free biscuits to the celebrants; portions of rabbit stew were paid for by Erie men to support various civic projects; the ladies ate their mulligan stew free of charge.

Late in May 1876, the prospering little town suffered its first major flood, which, in later years, the experts determined was a one hundred-year flood. The *Greeley Sun* headlined the story, "Coal Creek Inspired Terror on Monday Last, If Never Before":

> About noon the streets of Erie began to fill with water, and before an hour had elapsed, houses were flooded and the whole flat on which the town is located was swept by a muddy, roaring flood, over a quarter of a mile in width.[17]

Fortunately, no one died.

> People waded through the boiling flood with children on their shoulders, when the current threatened every moment to dash their feet from beneath them and to hurl them helpless and drowning down the roaring tide.[18]

The five-year-old tracks along the railroad line from Brighton washed out and required the laying of 1,555 feet of new line.

Charles Clark Welch of Golden, who found the first coal in the Louisville area.
—*Drawing,* History of Clear Creek and Boulder Valleys

CHAPTER THREE

While participation in the Erie Wars was a consuming occupation for some, others with solid financial backing were looking over the unexploited sections of the Northern Coal Field.

Charles Clark Welch, vice-president of the Colorado Central Railroad, was looking over a new area in August 1877 for possible new coal shafts. Together with his own leased 240 acres, he also secured the mineral rights to farmer David Kerr's wheat fields. (Kerr was to receive five cents per ton for any coal mined on his 150-acre property.)

Welch hired Louis Nawatny, a Polish immigrant and three-year resident of the area, to find a coal seam beneath Kerr's farm. Nawatny had no experience in locating coal banks, but he had been successful in finding underground water. Nawatny did find coal, and he constructed a vertical shaft two hundred feet below the Kerr wheat fields. The discovery developed into the Welch mine, which became the largest coal operation in Colorado at that time.

Nawatny turned to other business opportunities; he bought forty acres of homestead land nearby for ninety dollars, under a grant signed by President Rutherford B. Hayes, platted the land, named the area Louisville after himself, and proceeded to sell lots to incoming coal miners. He built a house for his family and saw to it that his wife Kathinka became the first postmistress of Louisville.

Nawatny's ambitious real estate program did not flourish, however. New miners had not yet saved enough money to buy the lots. Heavily in debt, Nawatny disappeared, apparently, in order to avoid his creditors.

But the new town of Louisville survived and grew to a population of two hundred by 1879, "made up of the Company, a few traders, employees, and railroad men."[1]

Louisville was a railroad town now that the Colorado Central had abandoned its earlier "snake route" to Boulder. By February 1878, the railroad had built a more direct grade from Golden to Louisville and on to Davidson and Boulder, a year later. The little coal community now planned for churches and a school.

The Welch mine was the center of the town's vitality, and a prosperous future seemed assured for the 175 miners employed there. One year later, owner C.C. Welch sold the mine—the *Rocky Mountain News* of September 30, 1879, reported he "secured a good round sum"—or was he forced to sell, to a slight, sallow little man with a wooly beard and white plug hat who represented the

A view of early Louisville's Spruce Street, looking toward the mountains.—*Photo, Louisville Public Library*

Union Pacific Railroad. This man was quietly buying up coal fields and railroad lines in Colorado and now had controlling interests in both the Kansas Pacific and the Colorado Central railroads. The man in the white hat was the notorious Jay Gould, who has been variously described as the "Mephistopheles of Wall Street," a "double-crosser, dishonest, amoral, and rapacious."[2]

Despite these "compliments," Gould was persistent and found more and more properties to add to the Coal Department of the Union Pacific. By 1880, through plot and counterplot, the financier had bought out the Colorado Central Railroad; the Welch mine was eventually renamed the Louisville mine and was wholly owned by railroad interests. This was not an isolated instance of absentee ownership, of course; mine after mine in the northern field came to be owned by someone who lived somewhere else.

At this time, William Davidson decided that running a coal mining company was not the life for him, and he sold out to concentrate on his agricultural pursuits.

To the south, the settlement at Marshall had been surveyed and platted. Joseph Marshall's long wait for the railroad was over. The Golden, Boulder and Caribou's engines steamed in on February 18, 1878, to move tons of Marshall coal to the markets. Marshall was able to double his production in one year; in 1878, before the railroad came in, he sent twenty-five thousand tons of coal to market; now, fifty thousand tons were loaded onto railroad cars.

Marshall had not lost his touch for innovation. He had developed a new device which was installed sometime before 1880:

Jay Gould confers with fellow railroad man W.A.H. Loveland.—*Photo, Denver Public Library*

> The coal is hauled out of a slope, and has hitherto been done by mules: but the Company are now putting in a double-engine, with cylinders nine inches in diameter, and a three-foot strike, the connecting rods being directly connected with the spool shaft. The spool is six and one-half feet in diameter, and will wind up 1,000 feet per minute, and can easily bring up six or eight cars at once.[3]

With progress often comes grumbling. Parents of Marshall children, who went to school in a one-room building between Marshall and Eldorado Springs, complained that the smoke and cinders from engines and locomotives clouded and befouled the children's playground.

Boys on one side, girls on another. Early school between Marshall and Eldorado Springs. *—Photo, A.A. Paddock Collection, Boulder Historical Society, Carnegie Branch Library for Local History*

Although Joseph Marshall's coal hauling machine replaced a few mules at that site, it would be many years before the animals were retired by machinery at most locations. In the slope mine, the mules could be brought out of the entries each evening to fresh grass and hay. That was mule foreman Joseph Borstnick's last job of the day at the Gorham.

In the shaft mines, however, the animals were stabled underground and seldom, if ever, came up to the surface until their working years were over. Sometimes the warm weather slack season permitted a summer vacation above ground. It was not unusual for an old mule to be totally blind when he came out of the mines.

Since these strong animals are the offspring of a horse and a jenny and can bear no young, mules have been characterized as "without pride of ancestry or hope of posterity."[4] Despite their ignoble background, most coal mine mules cost more than non-thoroughbred horses and were highly regarded by the mule foreman and his nippers—young assistants who were learning the trade. A man who would abuse a mule in a coal mine might well be blacklisted by the entire Colorado coal industry. Union officials, as well, would not defend such a man.

Despite all parties having no use for a mule abuser, "half the mules in the mines had their ears burned off"[5] from gas explosions caused by sparks given off by loaded cars on the tracks.

Each mule was carefully trained; the mule foreman got extra pay for "educating" a new mule. All had special names, were petted, and given special rations. August Figuerelli, mule boss at the Evans mine near Frederick, remembers, "I had 18-20 mules and I talked to them like pets."[6] Mules were the subject of many a poetic endearment:

> My sweetie's the mule in the mines.
> I drive her without any lines.
> On the rumble I sit
> And I chew and I spit
> All over my sweetie's behind.[7]

Stories of the mule's great intelligence were widespread. Mules seemed to know when it was time to quit for the day, and were considered "worse than a balky woman." Mules would pull two or three cars loaded with coal but would not move at four. Mules voted in some elections, according to those who criticized coal operators' methods of rigging the polls. In fact, each mine registered its mules. "They were unionized before some of us,"[8] claimed miners who reported that mules were more highly regarded than the men by the operators.

If the loaded coal cars descended a slope underground and then were tracked uphill, the mules would "beat the devil down hill so that they would get up the other hill without stopping."[9]

If car and mule reached an unsafe speed, either the mule driver or his nippers

A mule hauls coal.—*Photo, Colorado Historical Society*

Company houses and the boardinghouse at the Gorham mine, near Marshall.—*Photo, Louisville Public Library*

placed sprags* between the spokes of the wheels of the loaded cars to slow them down. The width of a passage was seldom much wider than the width of a coal car; there was no place to hide for the miner who turned to see a runaway coal car racing toward him. Furthermore, wooden sprags could break apart—another way a coal car could pick up enough speed to kill.

The mule driver directed his animals by simple "gee" and "haw" commands; he carried a whip but was loath to use it except in emergencies. Fire underground might require that the mules be brought up. When fire broke out in the Evans mine in 1918, the mules were "aproned" with leather straps or boards and brought to the top by the tipple's hoist and cable.

Mule drivers themselves were often the subject of "folk" stories. State Coal Mine Inspector Tom Allen said they "were reckless, indifferent men," but they had good qualities as well, "were moral and all that, excepting they drank and they spent their money, and just had one hell of a time."[10]

To be in good standing with the mule driver was very important to a coal miner. He was the one to supply empty coal cars underground; no topside foreman told the mule driver who got an empty car first unless the foreman was "down" on a particular miner. The mule driver might deliver cars for special favors—many agreed that an offered bottle of whiskey helped—then again, he might pass by the miner waiting to load coal.

Despite mechanization of the coal industry, mules were used in some mines until well into the 1950s. They were particularly valuable in gassy mines in which

* Wooden or iron bars the size of a small baseball bat with a crosspiece at one end.

electrical machinery might be a hazard. "Mules are more dependable and a damn sight cheaper than power . . . just shovel hay to them and they never stop."[11]

When Marshall gained a union in 1879, the first job for its leaders was to assess a strike call to the south at Como, Colorado. The astute northern miners, who worked Marshall's Black Diamond and Fox mines, realized that northern coal would be in great demand because of the strike to the south. So they called a strike as well. Within twenty-four hours, the Marshall strike was over. The northern operators, anxious to sell coal to the railroads, settled a wage dispute with the miners immediately.

Miners in the Canfield-Erie area also went on strike at this time, but with different results. The operators decided to break the strike with the first importation of Italian immigrants to the area. Shortly before Christmas, however, the Erie miners finally succeeded in forcing out the Italian workers, not without some violence, and the operators again reluctantly negotiated with their original crews.

Perhaps Joseph Marshall looked into the future and decided that labor troubles were not to his taste. Whatever the reason, late in 1879, he leased his 1,480 acres of coal banks for a ten-year period to his former partners in Black Hawk. Tubal Cain departed. Augustine Langford came down from the mountain camp to become general manager of the Marshall Coal Mining Company. His older brother, Nathaniel P. Langford, became president of the firm, probably in name only, for N.P. Langford was off shaping the history of Bozeman, Montana, and forming the boundaries of the nation's first natural area to be set aside, Yellowstone National Park.

In August 1878, the Knights of Labor tried again. This time, Erie miners W.E. McCoy and Ray Smith were successful in organizing sixty men in Local 771, Knights of Labor, the first national union in Colorado. Miners in the Lafayette area soon followed them with a local of their own; the two union groups represented about 5 percent of the men working in the Northern Coal Field.

That same year, thirty-seven-year-old Lafayette Miller died in Boulder of Bright's disease. His funeral service was well attended; the famous Erie Brass Band came in to provide a stirring requiem. The *Boulder County News* reported that "all possible respect and funeral honors were paid to the memory of the deceased. The Odd Fellows in regalia, and three fire companies of this city, and a concourse of citizens in carriages. . . .The Hook and Ladder truck was draped in mourning, with evergreen decorations."[12]

Miller's widow Mary left Boulder to turn her attention to the 1,280 acres of fenced farm land the family had accumulated east of Boulder. After several years of farm management, which resulted in luxuriant fields of oats, corn, wheat, and hay, hundreds of horses, cattle, and pigs, as well as extensive orchards, vineyards, and berry patches, Mary Miller was praised by the *Boulder County Herald* for her "well-conducted"[13] farm.

The newspaper went on to report that two years before, in 1884, a fourteen-foot vein of coal had been found on Willow Glen.

After some soul-searching, which included turning down an offer of $100,000 for the section of land containing the coal seam, the widow finally signed mineral rights releases, for a 12½-cent-per-ton royalty, to several mine promoters who began to strike shafts underground below her wheat, strawberries, and grapes.

Belle McDonald poses with some of her boarders at the Silver State Hotel on Front Street, 1874.—*Photo, Louisville Public Library*

CHAPTER FOUR

> Wheat-fields over coal mines, and over the wheat-fields, the charm of clear sky touched by the grandeur of the great mountains—can a spot of earth be better utilized, or be more beautiful for homes?[1]

Those who had bought Louisville lots from Louis Nawatny built modest homes as soon as they had saved enough from their wages at the Welch-Louisville mine. Small kitchen gardens began to green the area; their produce kept mining families going during the summer months, slack time at the mines.

By 1880, some six hundred residents had established roots in the new Louisville, their little houses clustered about the Colorado Central Railroad stop, earlier called Coal Creek Station. Most of the men were experienced miners whose northern European backgrounds are evident from the chart below.[2]

NATIONAL ORIGINS OF LOUISVILLE MINERS, 1880

England	43	Germany	7
United States	30	France	6
Austria	26	Wales	6
Scotland	21	Canada	3
Ireland	17	Russia	1
Nova Scotia	13	Switzerland	1

But now immigrants with French names and no coal mining experience had arrived. The new residents, and the Italians who would soon follow them in the 1890s, understood early on that they were to live apart from British and American families who had settled in a Louisville section called Kimbertown. ("Kimber" is a Cornish coal mining term.) Soon, separate Louisville neighborhoods were known as "Frenchtown" and "Little Italy," or more impolitely, "Dagotown." The Slavs also lived near one another. All of the newcomers lived east of the railroad tracks.

Front Street was becoming a solid row of thirteen or fourteen saloons, confined to that area by town ordinance. The management of one beer parlor decided that the pleasures of drinking could be profitably combined with the establishment of a bowling alley on the premises. Miners who came from similar national origins preferred to drink together as well as live near one another. The groups did "get together" from time to time in skirmishes and fistfights outside the saloons on Front Street.

Miners from eastern Europe started immigrating to Colorado in the 1890s.—*Photo, Colorado Historical Society*

The temperance movement was active in the coal towns as well; in 1881 Louisville celebrated the completion of the Temple of Honor for the Golden Lodge, whose purpose was eventually to close down the saloons, an uphill battle in any coal mining community.

Well before the town was official, young Louisville went to school, moving from one temporary location to another until a proper school building could be constructed. One wonders what happened one opening fall day in 1875 when Miss McCarthy, who had just signed a teaching contract for twenty-five dollars per month plus room and board, conducted classes for twenty-five Louisville youngsters in Jacobs's stone milkhouse until noon, at which point she resigned.

A two-story school building was started in 1880, built "out on the prairie," away from the center of town at Spruce and Main streets; by 1882, it housed 140 pupils. The building's first floor was used as a social center when school was not in session.

The community's first newspaper, *Louisville Miners*, started optimistically to print the news in 1887, but the firm died the following year.

Louisville Catholics built their church in 1884; it was the first religious building in town. Those who brought their Methodist Episcopal faith with them from Wales and Cornwall also supported a church that was completed in 1891.

As was common with emerging western towns, citizens had little interest in establishing sewers and other sanitary systems. Accordingly, during the summer and fall of 1881, Louisville suffered its first typhoid epidemic, which took the lives of a number of children.

Incorporation for Louisville was a slow process because some of the residents feared it, saying that the town "might be cleaned up . . . also whiskey would cost more."[3] Despite local opinion that government brings with it questionable progress, John H. Simpson was elected the town's first mayor in 1882.

More and more company-built residences—little box houses—dotted dusty streets. The company store flourished with the miners' enforced patronage. Above the store on the second floor, dancing parties were regularly held, many on payday. Despite the success of the company store, Louisville began to support a few privately owned groceries as well. In order to compete with the company store, the private firms offered credit during the summer months or when a strike was called.

As more and more company stores were established in the coal field, a deception was practiced that lasted well into the 1930s. A family's new household acquisition—curtains for the kitchen or a new sofa—was never admittedly

The Temperance Hall at Louisville, 1905. Tom Metz, proprietor.—*Photo, Louisville Public Library*

"bought" at the store competing with the company-controlled operation. An aunt or uncle had "given" the family the new chair or ice chest.

To celebrate Independence Day 1880, Louisville residents loaded up their wagons and buggies with bulging picnic baskets and freshly scrubbed children and urged their horses toward Dexter Grove, a park-like setting of cottonwoods a few miles outside of town along one of the area's irrigation ditches. Lawyer Alpheus Wright came out from Boulder to address the crowd. A performance of *Uncle Tom's Cabin* was a great success.

The following year, the Miners and Laborers Protective Association of Boulder and Weld Counties was born. Despite the imposing title, the group's main function was social; the officers planned dances, outings, lectures, and concerts for the benefit of the miners and their families.

C.C. Welch was still on hand in 1880 when the miners went on strike because the railroad promoter offered wages of $2.50 per ton rather than the usual $3 or

Pupils of all ages were educated at the Old Fairview School near Louisville.—*Photo, Louisville Public Library*

$3.50 per ton. Although the Louisville miners were novices at union activity, they were sophisticated enough to wait until the first cold fall day to voice their demands. Winter meant coal delivery. Welch capitulated in three days, no doubt fearing he would have to explain fleeting income figures to Jay Gould. Now Erie, Marshall, and Louisville had unions.

To the north, Canfield and Erie continued to flourish with the advent of yet another railroad. The Denver, Longmont and Northwestern line, known almost immediately as Longmont's "Baby Railroad," had laid 8.2 miles of narrow gauge track from Longmont to the Erie area. Both Isaac Canfield and William O. Wise were listed among the eleven names on the November 1878 incorporation papers. Up to this time, Marshall coal was hauled via the Colorado Central for Longmont's use. Erie and Canfield interests, however, wanted to corner the market with the little narrow gauge road, which began to haul coal from their mines by November 1881 at a lower rate.

A few weeks later, the "baby" line connected with a larger railroad, the Denver, Utah and Pacific, at the Mitchell stop. Now Longmont travelers had a shorter way to get to Denver. They paid a $2.50 fare; their horses, however, could be shipped for only forty cents.

Since excursions were the main entertainment of the 1880s, Longmont

A newer school for Louisville, built in the 1880s.—*J.B. Sturtevant photo, Louisville Public Library*

residents delighted in a trip on the baby railroad to Erie for a day at the skating carnival or a round of band concerts. Erie families could entrain for Longmont as well to enjoy Pumpkin Pie Day or stay overnight with friends for the annual Firemen's Masquerade Ball.

Prospects for the baby railroad dimmed when the Colorado Central, now part of Jay Gould's Union Pacific, retaliated with lowered excursion and shipping rates. The little Longmont line operated for eight years only, then was swallowed up by larger railroad companies. The narrow gauge tracks were torn out.

In 1887, John Simpson arranged a lease with Mary Miller to sink the first coal shaft on her land, down 240 feet to what would become the rich Simpson mine. Nearby, James Cannon also readied his coal mine for production, the mineral

The Methodist Episcopal Church, corner of Jefferson Avenue and Spruce Street, Louisville, 1892.—*Photo, Louisville Public Library*

rights for which he had also leased from the Miller family. The Cannon mine, however, did not flourish and was plagued from the beginning with quicksand cave-ins. It closed shortly after it opened. Soon, Joseph Mitchell sank a shaft on the McAffee farm nearby; within a few months, the Excelsior mine and the Spencer opened, along with the Gladstone mine, which started operations in 1890. Because of extensive mine expansion in the Lafayette area, the Colorado Central found it expedient to build a branch spur to the center of Mary Miller's lands.

Meanwhile, the enterprising farmer had other matters on her mind. Since her husband's death, she had been living with her six children in a modest cottage on the southwest corner of her property. In 1888, she was ready to build a house and found a town. She contributed 150 acres of her land, had them platted in January, and named the new community after the departed Lafayette. (For some reason, the town was replatted the following January by C.A. Russell.)

Mary's son Thomas J. Miller incorporated the town in 1890, the year he became the first mayor to Lafayette's 250 residents. By the end of 1891, Lafayette had "boomed" to 1,000 citizens and was surrounded by four active mines—the Simpson, the Mitchell, the Gladstone, and the Excelsior. Mrs. Miller sold more of her land to the expanding town.

The naming of Lafayette streets received Mrs. Miller's special attention: Geneseo Street was named for her birthplace in New York; Iowa Street commemorated the state where she married Lafayette; Cleveland Street honored President Grover Cleveland; Michigan Street got its name from Cleveland's running mate, A.G. Thurman, who was from Michigan (despite this celebration,

Patrons of the Louisville Beer Hall, 720 Front Street, before it burned down in 1897. — *Photo, Louisville Public Library*

the ticket lost that year); Couch Street (which is now Gough) memorialized the area's first physician; Emma Street was named for Mrs. Miller's favorite niece.

Her home on East Cleveland Street was one of several that were built in 1888. The building had to be large enough to house her immediate family as well as her parents and other relatives. The Miller home was the first in the area to include a well-appointed bath with inside water closet, bathtub, and lavatory, carefully inspected with awe by visitors to the house.

Her home established, Mary Miller started to think of a school, a church, and a general store. One establishment the businesswoman would not willingly foster, however, was the saloon. She was devoted to the temperance movement—later on, she was a candidate for state treasurer on the Prohibition party ticket—which had strong support in at least part of Boulder County. Those who bought lots from her had to promise in writing that no liquor would be sold on their land. Fifteen saloons did indeed flourish in Lafayette, but they did so out on the highway away from town.

The first company store, the Miners Trading Company, probably built in 1893, stood at Pine and Main streets until subsidence caused its condemnation.—*Photo, Louisville Public Library*

Mary Miller's business instincts were sound enough eventually to command a fortune of $250,000, an impressive sum for those days. She dressed the part—some called her Queen Elizabeth—and entertained frequently and graciously.

By 1889, Lafayette's first school building was completed; its doors were open to young boys only; later, girls were encouraged to attend as well. The schoolmaster was invited to board at the Miller residence.

Small elm trees began to shade the streets of downtown Lafayette. In 1892, a waterworks, which included a twelve-thousand-dollar reservoir and locally manufactured wooden pipes, was constructed.

Lafayette's founder also built the Congregational Church in 1892 and, with her usual largesse, paid the minister's salary for a number of years.

Not all Lafayette residents were Protestant. The Catholics, who did not have a church of their own until 1904, walked the four miles to Louisville for Sunday mass. No matter. Miners' families were used to walking to neighboring towns in those days. Moreover, Catholics were not made to feel welcome in Lafayette. Louisville residents did not feel welcome either when they visited Lafayette. When Lafayette suffered the first of periodic diphtheria epidemics in 1888, guards were posted around the town limits to keep Louisville residents out. When Louisville got sick, it returned the favor and posted guards to keep Lafayette out. The few black visitors to Lafayette were told that there was an ordinance in town that Negroes could not stay overnight; they were told to get out of town before dark.

Soon after Lafayette's founding, the town began to experience labor problems.

A social evening at Langford, April 6, 1889.—*Kneeland photo, Western Historical Collections, University of Colorado at Boulder*

In August 1889, alert members of the Northern Federation of Coal Miners called for Denver residents to boycott Lafayette coal shipments because most of the mines near town hired scab labor.

The decade of the 1880s was not a golden one for union development in the northern field. The rapid growth of the 1870s, despite an 1873 recession, did not continue into the 1880s. Striking miners did not have the clout they once had. More coal mines were opening in the southern field now, a region stretching from Walsenburg on the north, south to Trinidad, and on into New Mexico. The mines there were bigger, richer, with better quality coal, and soon their production outstripped that of the northern field, whose operations by comparison now seemed meager, even stunted.

In addition to the differences in coal production between southern and northern fields, the northern Colorado communities began to develop a special character distinct from the camps to the south. Historian Donald McClurg compares the two areas:

> In the north, the miners lived in small towns; towns which reflected the primarily rural character of the surrounding farm population, upon which the life of the town was in part based. Many of these northern miners came in time to be home-owners and part-time operators of small agricultural plots. In the south, the companies had erected camps at the mining sites, and these camps were surrounded by miles of barren plain, harsh and arid country.[4]

McClurg also comments on another special quality of the miners of the Northern Coal Field:

> There grew among the miners a tradition of north-south solidarity, based partly on the feeling that the northern miners owed a debt to their less fortunate southern brothers, and partly on the conviction that the welfare of the northern miners depended ultimately on the strength of organization in the south.[5]

For example, when southern coal miners struck in the fall of 1884, not only did the miners of Boulder, Weld, and El Paso counties participate in a sympathy strike of protest, but they contributed a portion of their wages to their southern brothers until the strike ended in December.

Wages did increase a bit in the north during the 1880s, but working conditions deteriorated to an even more dangerous degree. Those area residents not intimately connected with coal mining began to learn about the dangers of the occupation and the high accident rate. Some were shocked. Perhaps this outrage paved the way for the Colorado General Assembly's passage of a bill in 1883 authorizing the hiring of a state inspector of coal mines. The first inspector, John McNeil, did not mince words when he made his 1884 report to the governor:

> The coal mines of the State, as a rule, have hitherto been worked in a rude, miserable, and even reckless manner. Upon examination of many mines I could find no trace of an attempt to work after any system to suit ventilation or drainage; rooms and entries were worked into one another, and an unsufficiency of pillars left, causing a creep or crush of the entire strata over the coal, folding and breaking into fragments resulting in caves . . . it is almost impossible to circulate a sufficient current of pure air, thus entailing physical suffering upon the miner, and gradually undermining his constitution by being compelled to follow his occupation in an atmosphere laden with carbonic acid and other foul gases met with in coal mines.[6]

A year before, McNeil had issued this circular on the miners' plight:

I have found miners at work 200 feet ahead of the ventilating current. When I looked into this close and dirty hole, I could hear the miner's pick striking the coal, but at this distance I failed to see the miner, but knowing from experience if I went in there, I would find him, I ventured in a little over 100 feet in this densely thickened air, and I could just see a glimmering light, which on approaching closer, looked as if there were a great many lightning bugs around it.

When I arrived at the face of this place I found that the miner had his lamp hanging downward on the rib side to enable it to give more light. The air was almost stagnant to the feeling, and densely charged with coal dust, and in a few minutes I could feel the effects of carbonic acid gas, better known to the miners as black damp. This gas is given off from the lungs, from the combustion of the workmen's lamps, from the decaying timbers, and from the explosions of blasting powder. It also exudes from the fissures of the coal, and from the floor and roof of the mine. In its pure state, it is a most deadly poison, and will support neither life nor combustion.[7]

Earlier, in 1877, Colorado's General Assembly passed a law stating that young men could not work in the mines until they had reached their fourteenth birthday.

This charcoal drawing by Steinlen illustrates John McNeil's first reports to the governor.—*Drawing, Western Historical Collections, University of Colorado at Boulder*

The lawmakers knew that boys as young as five years old were often hired as breakers or trapper boys. When John McNeil visited both southern and northern areas a few years later, he still found nine- and ten-year-old boys working underground, despite the law. McNeil revised the minimum age law in this way: "No young person under 12 years of age, or woman or girl of any age, shall be permitted to enter any coal mine to work therein, nor any miner, under the age of 16 years, unless he can read and write."

Even the earliest coal miner was dependent upon the work of others. The mule driver brought him empty cars and took away those he had loaded. The fire boss went into the mine before each shift with a safety lamp to check for gas pockets. (If the lamp's light flared high, an amount of explosive methane gas was present. If the flame went low or went out, the fire boss knew that black damp, carbon dioxide, was present and he got out fast.)

Outside, the weigh boss stood by the tipple as the car unloaded the coal into the large freight cars to determine the miner's tonnage, by which he was paid. (The miner hooked his brass or iron check to each car he loaded.) The miner, of course, was underground and unable to know whether or not he was being treated fairly. For years, miners had suspected, sometimes proved, that company-hired weigh bosses were less than scrupulous about weighing the coal correctly. Their "mistakes" never seemed to favor the miner. Most miners felt they should be able to select their own checkweighman who would watch to guarantee that the company was not skimming each each coal miner's tonnage to pay for operating the mines, as some alleged.

Despite this interdependence among coal workers, the miner regarded himself as an independent contractor. He brought to the job his expertise. He purchased his own equipment—shovels, picks, copper needles, tampens—and paid for its repair. He bought the black powder that he injected into the coal seams to release the chunks; later, he paid for the dynamite he used. He supplied the lard oil and wick for the lamp he wore attached to his soft cap. (By the 1920s, most miners used a hard leather cap.) When his head lamp was lit, the flame came out of a long spout. The mules wore a smaller version of the spouted lamp on their harnesses. (Whale oil was used for this type of lamp in the old country.)

By 1910 or so, some miners replaced the wick lamp with a carbide light, which was powered by a combination of calcium carbide and water to produce a white light. Now and again, the carbide light would flash when the wearer came near a pocket of gas; the resulting smell was similar to expended firecrackers on the Fourth of July.

The miner's lunch bucket was made of iron graniteware; the top section held his lunch with maybe a special shelf for a piece of pie; the bottom of the pail contained water, an important supply to the man who might be temporarily trapped underground. The heavier granite lunch pail was replaced by one of aluminum sometime after the turn of the century.

The coal miner was proud to be his own man, a skilled laborer who contracted with the operator for the job. The operator also regarded the miner as a contractor, a man who was paid for the tons of coal he mined, a man who was not paid for such chores as safety timbering of each room and entry, screening of coal, removal of slag and rock piles, or the laying of railroad track. This was the "deadwork" for which the miner felt he should also be paid. The operator felt that, as an independent contractor, the miner should bear the expense of deadwork. (Miners were not paid for any form of deadwork until the 1920s; at that point, the amount was minimal enough to call for a strike.)

In truth, the coal miner was a wage earner and not a contractor, as he saw himself, and as the operator continued to view him long into the twentieth century. To come west to Colorado was to come to the frontier of independent men—the prospector, the gambler, the mountain scout, the Indian, the gunfighter. Perhaps the coal miner did not realize he was participating in another frontier as well, not as romantic, but the West was becoming the "wageworkers' frontier"[8] and would add an urban tradition to its usual theme of independence of action. Since the coal that the miner supplied to the operator was subject to the economic laws of supply and demand, he was often helpless to control his employment situation. Most miners were beginning to see that affiliation with a large, strong national union was a necessity.

Whenever miners had the chance to meet together, most agreed that, in addition to an employee-selected checkweighman and proper pay for deadwork, they

Timbering at the Monarch mine.—*Photo, Colorado Historical Society*

wanted to be paid twice per month, in gold or by check, not with scrip, a paper currency redeemable only at the company store, sometimes at a 25 percent discount. The issuing of scrip was just coming into vogue with coal operators, and the men hated it for what it was, a further restriction to their economic freedom. A fee for the company doctor was deducted from the miners' wages; another charge for the bathhouse was deducted in the same manner.

The men saw no reason why they should be forced to work underground for longer than eight hours per shift. They, like other people, wanted to feel the warmth of sunshine and to play with their children before bedtime.

Colorado farmers, a powerful conservative force in state politics, could not see the sense of an eight-hour day; after all, they said, farmers could not work that way, why should the miners? Because the miner was paid by the ton of coal he extracted from the ground, and not by the hour, he had difficulty convincing the Colorado public, as well as the coal operator, that eight hours were enough to spend underground.

Oh, they had the eight-hour day, all right, joked one miner, "eight hours before dinner and eight hours after."[9] Coal operators publicly agreed with the miners that the eight-hour day should become state law, but privately they ignored the concept. The operator knew that if a miner did not want to work more than eight hours per shift, he could be replaced with someone who would. And the miner knew it as well.

On November 1, 1882, five hundred miners walked off the job at the Star and Mitchell mines. Through their spokesman, G.T. Bailey, they demanded more money for such hazardous employment. The next day, more than two hundred miners at the Stewart mine north of Louisville followed suit. Eventually the strike was to involve twelve mines. The men were earning eighty cents per ton and they wanted one dollar per ton. Strikebreakers were brought in. By the end of November, the Louisville strikers got a ten-cent raise, but the Canfield miners cooled their heels until January 1883, when their demands were finally met. Had these local union groups worked together more effectively, or had they been affiliated with a strong national union, perhaps both crews would have been back to work sooner. In 1883, another short strike affected all mines in the northern field. The next year, two northern mines closed for a time in a dispute over the hiring of a checkweighman. The men went back to work without solving the weighing problem.

Meanwhile, Knights of Labor organizers continued to work Colorado coal fields for membership and improved working conditions. Agents John R. Buchanan, John L. Lewis (not related to the later president of the U.M.W.A.), and Charles J. Driscoll were active in El Paso County during the fall of 1884. Buchanan advised the miners there to wait for the first snowfall before presenting their demands. Snow fell on October 27; twenty-four hours later, with back orders for coal delivery, the operator was forced to negotiate with the El Paso County miners.

That same year, Joseph Marshall experimented with another new device, an undercutting machine that involved three mechanical picks. The miners used it reluctantly because they said that they were deafened by the noise of the machine and, therefore, could not hear the warning noises the rocks made just before a cave-in.

More and more coal fields in the northern field had been bought or leased by eastern financial interests. The miner seldom had the opportunity to face the mine owner; he lived somewhere else. By 1885, Isaac Canfield's old Rob Roy mine was owned out-of-state; the same was true of the Star and the Mitchell. Local managers seldom had a financial stake in the companies they served. They stuck to the rules laid down by eastern owners, the most important of which was never to recognize a national union as a bargaining agent for the miners.

The original section of the Boulder Valley mine, long the focal point for the Erie Wars, closed in 1886; development of its underground chambers had been faulty enough so that the mine could not remain open safely.

A new state organ, the Bureau of Labor Statistics, issued its first report in 1888 and had this to say about the life of a coal miner:

> There is, perhaps, no class of laborers more deserving of assistance and sympathy than coal miners. The nature of their employment isolates them from their fellow-workmen during hours of labor which are usually so long as to leave but little time for self-improvement or for the enjoyment of such rational amusements as "make life worth living." Employers, knowing their condition, take advantage of, and many times impose burdens upon them that seem outrageous to contemplate.[10]

In 1887, a miner in Boulder or Weld County could earn $21 for a sixty-hour week. The young boys who helped them, albeit illegally, got $6 for the same long hours per week. Mule drivers received $18; firemen brought home $15. Weigh bosses and engineers earned $18, but they worked a seventy-hour week for that pay. The foremen worked the extra hours also and earned from $21 to $24 per week.[11]

Coal operators had imported Chinese scabs to Wyoming mines, most notably in Rock Springs. Miners in the Colorado fields feared the possible influx of Chinese and black work crews because they knew, as did the operators, that desperate men will work longer hours and for less money. Such crews were brought into the southern field when miners there complained of their long hours.

In November 1885, during a twenty-one-day strike, violence broke out near the Welch-Louisville mine when the men learned that two labor contractors were openly encouraging newcomers to scab for $3 per day. The regular miners were then earning about $4 per day, which amounted to 93½ cents per ton. Toward the end of the strike, a small group of strikers hijacked a company locomotive, tied up the engineer, and set buildings and piles of coal on fire near one of the Marshall mines. Even so, the men settled with the operator and received back wages.

Because of such incidents, the Colorado public expressed outrage at the miners' actions. Their original sympathy for the plight of the underground worker changed to disgust with the reporting of increasing violence on mining property.

By 1887, despite the lessening regard in which the Knights of Labor were held, local 9795 was organized at Langford. The community at Marshall had been officially called Langford since 1882 when Augustine G. Langford assumed management of the Marshall mines. Although Langford died in 1884, the post office was changed to bear his name; stubborn local residents, however, continued to call the settlement "Marshall," or by its nickname, "Foxtown." As they had in

Louisville, incoming French and Italian immigrants learned early that they were to live separately from the others and to patronize their own saloons.

Union men in the northern field began to argue among themselves as to the proper method to deal with continuing labor problems. In 1886, a new group formed in Louisville, the Coal Miners' Federation of Northern Colorado, whose goal was better negotiations between labor and management. The Federation's message was that individual union activity was no longer effective. Representatives from the Federation went to Denver regularly to lobby the General Assembly. They must have found a way to influence a number of legislators, for, two years later, a new state law made it illegal to blacklist miners and provided for fines for those operators who continued to hire youngsters who had not yet reached their fourteenth birthday. Even though such legislation was on the statute books, the coal operators paid little attention to the new laws.

The Bureau of Labor Statistics reported that "such has been the success of the Federation and with such moderation have its affairs been conducted that very few, if any, troubles have occurred between employers and employees since its formation."[12] Federation tactics seemed more palatable compared to those of the failing Knights of Labor. The two groups tried to work out their differences by meeting together on a common board, the Progressive Union of Miners and Mine Laborers, which formed early in December 1888.

One year later, another group was moving to power in the eastern states. Mining men there met just before Christmas in 1889 to form the United Mine Workers of America. It took just one month for the old Knights of Labor local 771 at Erie to join with the U.M.W.A. The local's fifty-five members quickly grew to ninety. Other Colorado miners' groups followed suit.

Association with the U.M.W.A. had its benefits for local miners, but there were drawbacks as well. Colorado, in general, did not favor large national organizations. Citizens were suspicious of decisions made by unknown union men in a distant office back east. Surely, Coloradans reasoned, miners and management could sit down and work out their differences at each mine. They overlooked the fact that local managers seldom owned any stock in the companies they worked for. The profit motive meant less to them; they were less tractable than those who owned the mines. Eighty percent of the coal mined in Colorado now went to fire locomotives whose lines were owned out-of-state. Railroad companies, however, were getting out of the coal mining business. One by one, the coal lands acquired by railroad firms just a few years before were sold or leased to new business combinations.

In 1890, one of these new businesses, the Rocky Mountain Fuel Company, was buying up a number of the mines in the northern field and would become an important force in labor relations after the turn of the century.

Delivering Lafayette's mail.—*Photo, Lafayette Public Library*

CHAPTER FIVE

> It is only a question of time when this immigration octopus will develop so many tentacles that it will reach into every avenue of labor, crushing the American standard of wages, and creating a surplus of workmen that will far exceed the demand.[1]

In the beginning of the 1890s, a curious organization surfaced in Denver whose tactics increased and coalesced Colorado's fear of further immigration. The American Protective Association, formed in 1887 in Iowa, felt its duty was to alert Americans to the continuing threats of immigration, which would cause economic hard times for all.

The organization grew to a Colorado membership of three thousand in 1893, then mushroomed to ten thousand stalwarts the following year with twenty-five chapters in the Colorado area. Although the A.P.A. was anti-Catholic, it did sponsor chapters for black Coloradans—separate ones, of course. The group worked its way into the Republican party and controlled its policies for a time, just as the Ku Klux Klan was to do later in the 1920s.

The *Rocky Mountain News*, populist in some of its editorial stands, characterized the A.P.A. as a group that included a few "dead-in-earnest crazy fanatics," some "unprincipled office seekers," and a "very large number of dupes."[2]

Some of the earliest miners, whose families had come from England, Wales, Scotland, and Germany, were attracted to the American Protective Association; immigrant labor from southern and eastern European countries had already displaced some of their relatives in coal mines back east.

Considering the large appeal of the A.P.A., it is easy to see why many Coloradans were suspicious of anyone who spoke a foreign language, had old-world ways, or whose skin was Mediterranean-swarthy. (When the Ku Klux Klan surfaced in 1915 as a dominant political power in the state for the next ten years or so, those who had held membership in the A.P.A. immediately joined the Klan. It was their kind of organization.)

Miners in both the northern and southern fields suffered from the activities of these radical groups.[3] No longer were the coal mines the exclusive province of the northern European and American as shown by the following chart:[4]

Foreign-born Population of Five Coal Producing Counties of Colorado by Place of Birth—1870 to 1910

PLACE OF BIRTH

	Foreign-born		Gt. Britain		Northern Europe		Southern Europe and Mexico		Other and Not Reported	
Year	Number	% of Total Pop.	No.	% of Total Pop.	No.	% of Total Pop.	No.	% of Total Pop.	No.	% of Total Pop.
1870	952	9.3	448	4.4	350	3.4	141	1.3	13	0.2
1880	3,944	13.5	2,600	8.9	815	2.8	140	0.5	389	1.3
1890	10,199	20.9	5,606	11.5	2,679	5.5	1,840	3.8	74	0.1
1900	12,873	18.0	5,302	7.4	3,374	4.7	4,042	5.7	155	0.2
1910	24,773	22.5	5,631	5.1	6,555	6.0	11,767	10.7	820	0.7

In the 1880s, four-fifths of the coal miners in Colroado spoke English. In the space of thirty years, the ratio of English-speaking to non-English-speaking miners gradually reversed. By 1910, 70 percent of Colorado's coal miners were foreign-born, spoke little English and had no previous experience in the mines. Now the state published its coal mining laws in Japanese, Italian, Polish, Hungarian, German, Croatian, Greek, Spanish, Bulgarian, and Russian, as well as English.

Many had been directly recruited in southern and eastern Europe by railroad lines, steamship companies, and coal mine owners. Ads were placed in old–country newspapers testifying to the great fortunes to be made in coal fields throughout the West. Earlier immigrants wrote home to say that the chance for the good life in America was far greater than for those who remained in countries where economies were breaking apart. When John Simpson sank the shaft on Mary Miller's land, he wrote home to England saying that more men were needed for his mine. As one miner who did come later reported, Simpson "neglected to mention they would be scabs."[5]

Recruitment was systematic and included ads "placed by steamship agents who traveled to the remote mountain villages to secure clients for their ships leaving Greece and other European ports for America."[6]

Sad to say, sometimes countryman preyed upon countryman, as in the case of Leonidas Skliris, an immigrant who enjoyed the title "Czar of the Greeks." Skliris had come to the United States in 1895 and wasted little time becoming a "labor agent" for Greeks wishing to work western coal mines either in Colorado or Utah. First appreciated by the newcomer, then hated and feared, Skliris's success and that of other labor agents depended upon the immigrants' ignorance of American ways. Quickly they learned it was not always necessary to pay to get a job in the mines.

Mine operators "encouraged" friendly newspaper editors to write glowing articles about the opportunities for work in the coal mines and elsewhere. The practice was common enough to merit discussion in the *Second Biennial Report of the Bureau of Labor Statistics, 1889–1890* which headlined an article, "Deceiving Workmen":

> A pernicious practice is indulged in by some newspapers and real estate "boomers" in Colorado, in making statements to the public that men are in great demand, for this or that vocation.[7]

In the *Lake City Journal*, August 12, 1890, a page-one article stated:

> Men for the mines are in great demand and work is suspended at some of the mines while operators are looking in vain for miners. There is work for several hundred miners at good wages.[8]

After investigation, the state labor agency reported that a Lake City miner said:

> The item is a wilful falsehood; of the old hands and old timers, a good many are idle at present. The idea is to flood the country with working men in order to get their labor for little or nothing.[9]

Certainly those who worked in the Northern Coal Field knew their jobs were on the line because of the operators' easy access to the new immigrants. On May 8, 1891, Marshall mine workers at Langford went on strike as a protest against the introduction of "a certain class of foreign labor."[10] The miners had help from an unusual quarter.

> A romantic feature of the strike was the fact that the wives of some of the strikers assisted their husbands and the strikers in taking possession of the mine. Armed with clubs and stones, this combined force successfully drove the imported men from the mines when they attempted to take the places of the strikers.[11]

By May 23, the women were able to attend to domestic matters again; their men went back to work at the Marshall #5 mine.

It was not only the influx of Italians, Greeks, Croatians, Russians, and Poles into the northern field that caused resident miners to reflect again upon their working conditions. The eight-hour day was on the state statute books now and again, but the men knew they had to work a ten- to twelve-hour day to keep their jobs. In 1892, Louisville and Lafayette miners struck for seven days to complain of their twelve-hour work day and a cut in wages as well. The strike failed. In 1899, the Colorado General Assembly passed an eight-hour-day law, but the Colorado Supreme Court declared that the legislation was unconstitutional. No legal body, the court reasoned, could dictate the daily hours a laboring man might work.

For years, unions and operators had signed agreements for checkweighmen to be chosen by the miners; seldom were these agreements adhered to. In 1897, miner-chosen checkweighmen were mandatory by Colorado law, but operators disregarded it.

Payment of paper scrip for wages, rather than U.S. currency, became common practice during the 1890s. At a mass meeting on October 25, 1893, Lafayette miners decided to strike because local companies were forcing miners to accept scrip as wages. The strike, scheduled for November 1, lasted only one day; operators capitulated and began to pay the men with proper currency—for a time. Evidently, the *Denver Republican* thought the whole affair ridiculous and headlined their story "They Insist on Having Money."[12] During the 1901 General Assembly, a no-scrip law was passed, another statute for operators to ignore.

Colorado's labor force of coal miners was growing. By the mid-1890s, 6,164 men worked the mines in both the northern and southern coal fields. In Boulder County, eighteen mines employed 602 men; in Weld County, three mines were open and employed 40 men.

In June 1894, a general coal mine strike throughout Colorado was called. Of special concern to miners were lower wages, issuance of scrip, and the further deterioration of working conditions in the underground chambers. They also resented the presence of company stores and the operators' continued thwarting of what the miners regarded as their unalienable right to organize.

The general strike failed. Scrip continued to be issued, the company stores flourished with enforced patronage and higher prices, and the operators threatened the miners again with scab labor that was waiting in nearby Denver to be brought by railroad boxcar into the field.

Despite reports to the governor from state coal inspectors, safety features were installed in only a few of the mines. A number had no ventilation system, the chambers were supported by inadequate timbering, and many mines did not possess a second shaft or sloped opening for quick exit in case of trouble.

A downtown building block in Lafayette, 1892.—*Photo, Lafayette Public Library*

Miners still paid for their tools and their repair; they still bought their own blasting powder and other supplies. Most accidents underground were ascribed to the carelessness of the miners when official reports were completed. Since the coal companies were seldom blamed for any accident, the widows of those killed in the mines had no claim upon them. If a miner was injured, he was examined by the company physician; he did not choose his own doctor. In view of the generalized lack of attention to the miners' workplace, it was not surprising that the death rate in Colorado coal mines was more than twice the death rate in coal mines throughout the nation.

So sporadic strikes continued. On May 18, 1895, miners at the Marshall mines struck to protest payment of wages with scrip and the lack of regular paydays. The operators settled with the miners within forty-eight hours. On May 22, one hundred miners at the Acme went out because they had not been paid at all for some time. Finally, the operator capitulated and paid the men. On January 18, 1896, Rex miners struck for the same reason and finally got two months' back pay. Three months later, miners from Erie, Marshall, and Lafayette met with Louisville men to discuss their mutual problems. A committee was formed to deal with each operator. The United Coal Company, the Marshall Coal Company, and the People's Coal Company would not cooperate with the committee. With a strike call for April 6, however, Marshall Coal and United Coal decided to go along with the miners' group.

During this time, the Western Federation of Miners enjoyed a brief period of popularity among some of the northern field miners. When United Coal wanted its employees to work on May 22, the date of the annual miners' picnic, L.W. Rogers of the Western Federation stepped in when it was learned that United Coal planned to use scab labor on picnic day and thereafter.

Virtually all of the men in the northern field struck to protest this outrage—Long's Peak mine near Erie stayed open—with the result that, on June 18, 1896, the Western Federation of Miners and most of the operators in the northern field signed a labor agreement. In 1897, the General Assembly passed a law which provided that a checkweighman "may be employed . . . who shall be selected by the miners employed in said mine and whose wages shall be paid by the miners therein employed."[13]

Superior farmer William Hake finally decided in 1895 to exploit the coal seam seam on his property, uncovered years before during the flood of 1864. Hake never took an active role in mine management but, instead, signed an agreement with James H. Hood to construct the primary shaft of what would become the Industrial mine. Hood had already opened the Hecla and would later sink shafts for the Centennial and the Vulcan mines. He became president of the Industrial Mining Company; around the turn of the century he sold the Hake family interest in the mine to the Northern Coal and Coke Company, which was to be absorbed by the growing Rocky Mountain Fuel Company in 1911.

The Industrial mine employed three hundred miners before mechanization; by the time it closed in 1945, almost four million tons of coal had been taken from its chambers.

The little town of Superior was platted in 1895 by the elder Hake. The town was surrounded by coal mines. In addition to the Industrial, the Enterprise opened nearby. The Monarch mine, southeast of Superior, opened in 1902. Years later, the Pluto, sometimes called the Red Ash mine, began operation. The *Boulder City Directory* for 1896 listed two hundred Superior residents and described the town as taking "a gentle slope toward the rising of the sun."

With the opening of the coal mines, other businesses came to Superior. Schilling House, a boardinghouse catering to unmarried miners, opened its doors, as did a general store and a candy and fruit shop. By 1904, the year of its incorporation, Superior supported two hotels, several saloons, and the Miners' Trading Company, which sold meat, groceries, dry goods, and clothing—for scrip.

Superior children had gone to school near town since 1901. Eight grades were housed in a little one-room building. Most Superior families considered the completion of the eighth grade an adequate formal education.

Fire protection consisted of two hand–pulled carts, acquired in 1904. Trains along the Union Pacific route occasionally stopped at Superior. The Billings Flyer came whistling through town at 9 A.M. from Montana and again at 7 P.M. on its return trip to Billings, but the Flyer did not stop at little Superior.

On the road between Superior and Marshall, a house was occupied by several young women who "made themselves available" for a fee. Apparently, business was brisk.

For several years before the turn of the century, small power plants operated in the Lafayette-Louisville area. They were modest facilities and did not extend service any great distance. Erie had its own plant.

On July 1, 1898, however, a headline in the *Denver Times* read: "Mammoth Power Plant to Supply Four Counties"; the article announced the incorporation of

Mayor Hake's Superior home.—*Sturtevant and Pruden photo, Boulder Historical Society*

The new town of Superior. The tall white building is the miners' boardinghouse.—*Photo, Louisville Public Library*

More company houses go up at the Industrial mine, Superior.—*Photo, Louisville Public Library*

the Rocky Mountain Power Company, whose forthcoming plant would be the "largest between Chicago and Ogden." "Within a week," the *Times* went on to predict, "work upon one of the most important enterprises ever launched in the West will have been commenced. . . .It is the intention of the company to furnish light and power to all of the mines, coal and metalliferous, within a radius of thirty miles of Lafayette. For the gold, silver, and copper mines, wires will be strung into Boulder, Gilpin and Clear Creek counties, and the entire coal field of Northern Colorado will be supplied."

The proposed power plant was slow to develop but, when it was fully operable by 1906, it did indeed fulfill most of the *Times*'s predictions.

Full use of electricity came slowly. In 1899, the Louisville Town Hall boasted two electric lights. Within the next several years, most homes had at least one electric fixture. There being no separate circuits at first, all the lights in the area went on at dusk and, after a few warning blinks, went off at 10 P.M. On dance nights, Lafayette and Louisville young people would try to bribe the night man at the "light plant" from time to time so that the dancing might go on for a little while longer. Sometimes, the bribe was successful.

By March 1900, Louisville had set out nineteen street lights. By 1904, morning electricity became available from 5 A.M. to first light. Nobody turned the light switch on and off. Subscribers left their fixture "on" all the time to take advantage of the first minutes of electric light in the morning and at night. Each fixture came

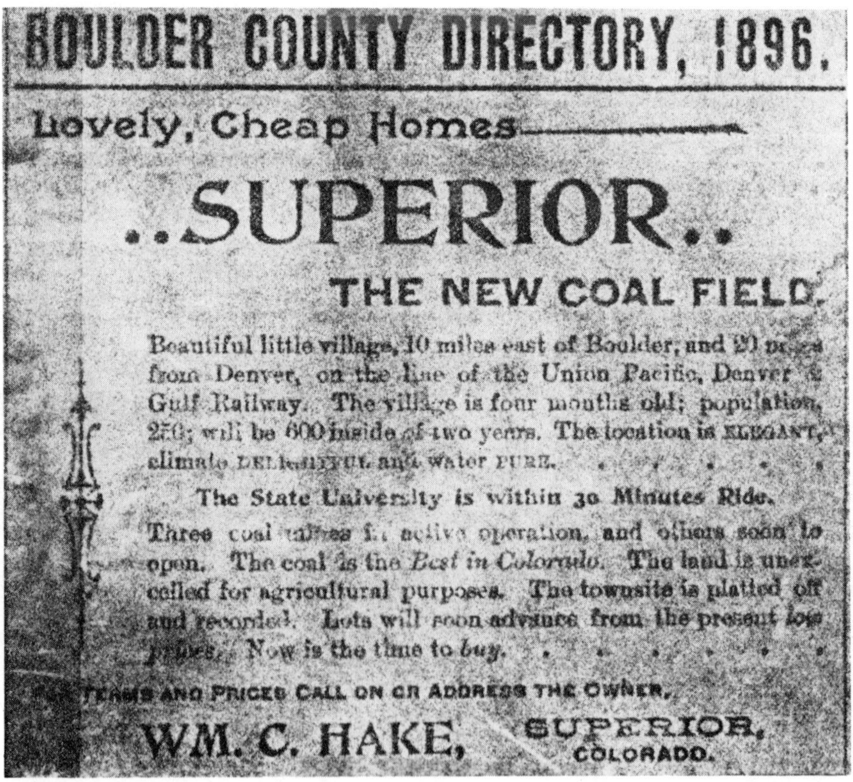

The company store at Superior.—*Photo, Louisville Public Library*

A country school near Superior.—*Photo, Boulder Historical Society*

The Industrial mine at Superior.—*Photo, Lafayette Public Library*

Western Power and Light Company, near Lafayette, the first such installation in the area.
—*Photo, Lafayette Public Library*

with a platinum bulb; when it burned out, the subscriber took it to the plant and received a new bulb free of charge.

By now, some fifty thousand dollars had been invested in the plant. Power was supplied by two 1,000-kilowatt, 25-cycle generators with 2,000 horsepower from the Rocky Mountain Power Company. A 60-cycle generator was eventually installed for domestic use. The coal used to power the facility came from a nearby mine named, appropriately enough, the Electric.

Some miners called electricity "white coal" and were understandably nervous about its use underground. Would it "spark off" gas pockets to explode below? Coal Mine Inspector Henry Denman was also concerned:

> In my biennial report for 1889-1890, I recommended several revisions that are much needed, but for some unanswerable reason the present administration ignored the same. Electricity is now being used extensively both for mining and hauling of coal. As yet there is no law made to compel insulation of wires or to prevent the use of electricity in gaseous mines. Apparently no consideration is given to the protection of Colorado's most important producers, viz.: coal miners.[14]

In 1906, the light plant, now called the Northern Colorado Power Company, offered service to Boulder, Longmont, Berthoud, Loveland, Fort Collins, and Greeley, as well as the coal communities in the area. (In 1924, the Public Service Company of Colorado took over the last of the state's independent power companies. The Lafayette plant was in full operation until 1929, and was in partial use until 1948, when Public Service shut it down for good. The building itself was razed in 1952. By then, Public Service had built a large plant east of Boulder, at the edge of Weisenhorn Lake and Amusement Park. The area came to be called "White City" by local coal miners.)

When the Colorado and Southern's Denver and Interurban trains began "kite route" service on June 23, 1908, from Denver "along the foothills and into the mountains," the cars were powered by the light plant near Lafayette. Sixteen cars went back and forth on the route each day; each car required eleven thousand volts for its power. These suburban trains enjoyed an enthusiastic patronage in the coal field for eighteen years. The cars—sometimes called pollywogs—were clean and ran on time. The trip to Denver from Boulder took one hour and eleven minutes. A spur line ran to nearby Eldorado Springs, now a fancy resort with hot springs, three swimming pools, dance halls, a skating rink, and a variety of hotel accommodations.

The train changed life in the northern field in a number of ways. Young people could commute to another town for advanced education; some came in to Boulder to study at the university. Farmers could pack a suitcase with fresh butter to sell in town. Refrigerator cars not only brought fresh produce to the coal towns, but also carried cold wooden kegs of Zang's or Tivoli beer. Passengers entrained for baseball games, movies in Denver, or special expeditions. The sick could be placed on the train for better medical care in town. It was wonderful while it lasted; as the automobile became popular, however, the railroad began to lose its customers; the Denver and Interurban closed down in 1926.

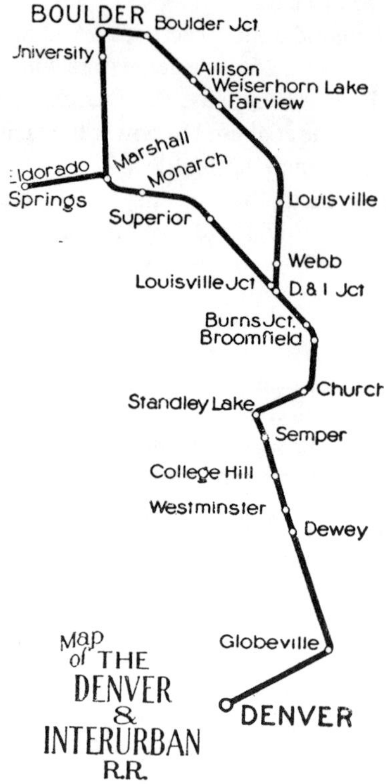

The Denver and Interurban tracks, looking north from the Louisville station, sometime after 1905. Note third rail, left track, for narrow gauge railroad.—*Photo, A.A. Paddock Collection, Boulder Historical Society, Carnegie Branch Library for Local History*

The Denver and Interurban begins its run.—*Photo, A.A. Paddock Collection, Boulder Historical Society, Carnegie Branch Library for Local History*

Maude Alice Machin is Lafayette's first telephone operator, 1902.—*Photo, Lafayette Public Library*

CHAPTER SIX

The little mining towns of the Northern Coal Field greeted the twentieth century with more street lights, the telephone, ice cream cones, and soda pop.

The forty-four street lights in Louisville gave the town a settled look despite its dusty, unpaved roads. The height of sophistication, telephone service, came to Lafayette in July 1903 with M.F. Madison as owner and manager and to Louisville with Martin J. Zurich as manager of the independent company. Sixteen-year-old Maude Machin became Louisville's first telephone operator, the start of a lifelong career with telephones. As business picked up, young Ambrose Taylor was hired as the night man. Three years later, 119 Louisville residents had telephones; Lafayette had 112. Both private companies were later bought out by the Colorado Telephone Company.

By 1910, almost everyone in the northern field could call the local operator for the correct time, a prediction of weather conditions, who was downtown, what stores were open, and who won the baseball game. A seemingly innocent device with few political implications, the telephone would play an unusual role in the coal strikes that were to come in the area.

By Labor Day 1901, the United Mine Workers had finished a union hall in Lafayette. Opening celebration included entertainment by Irish tenors, Welsh orators, play readings, and concerts by local brass bands who came out for both happy and sad occasions, for picnics and for funerals. (Some wit observed that playing in a cornet band was a healthy pastime for coal miners; the blowing forced the week's coal dust from their lungs.)

For special evening "balls," Stobb's orchestra entertained the dancers with concertina, piano, clarinet, and slide trombone. The Rex and the Model theaters opened in Louisville with silent movies; admission was five cents, then rose to an outrageous ten cents. By 1905, ice cream cones were no longer a novelty; the buyer could choose vanilla or strawberry. Soda pop became available in lemon or strawberry flavors.

Unlike the camps in the southern field, the presence of a company store did not stop the establishment of many small groceries in the area. The force of competition, however, encouraged grocers to call upon Louisville and Lafayette housewives each morning for their orders, the merchandise to be delivered that afternoon by horse and buggy. Many stores closed on the days baseball teams met

Another view of Louisville's Main Street. Town hall at left.—*Photo, Louisville Public Library*

Local 1388, U.M.W.A. hall in Lafayette at 209 East Simpson.—*Photo, Lafayette Public Library*

Louisville's public school for elementary children opened its doors in 1903.—*Photo, Louisville Public Library*

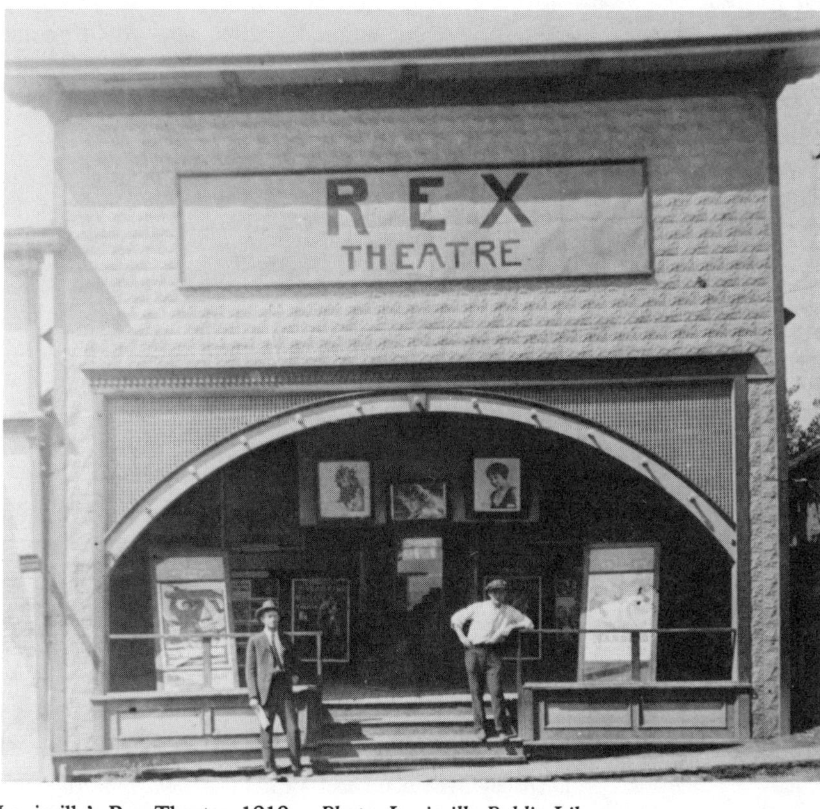

Louisville's Rex Theatre, 1919.—*Photo, Louisville Public Library*

to compete; Louisville and Lafayette were strong rivals. Baseball was also taken seriously in Superior, Marshall, Canfield, Erie, and Davidson.

Lafayette House, an elegant boardinghouse by 1900 standards, offered room and board for fifty cents per night, or three dollars by the week. Some of the unmarried miners could afford to stay there. Seldom was there a vacancy in a boardinghouse or private home.

Louisville's Taylor Stable was without management in 1902 when Mr. Taylor, who had deserted Mrs. Taylor for the pleasures of Lafayette—she managed the stable now—returned to ask his estranged wife for more money. Her stable was doing better than his new one in Lafayette. She said no; he shot her and then killed himself. Janet Taylor was buried under a tombstone that stated, "The world seems quite another place without the sight of Mother's face."[1]

Mary Miller now spent her time developing a new enterprise—a family-run bank. She had established a modest Miller Bank in 1892, which expanded into the Lafayette Bank in 1902. When she became president of the bank, proud Lafayette residents boasted that she was the first female bank president in the United States, perhaps in the world!

Mary's son George was vice-president, her son James was cashier, and James's wife Elizabeth became assistant cashier. (Mary's son Thomas J. Miller, who had served as Lafayette's first mayor, died in a Strathmore mine accident in 1902.)

Those Louisville residents who could read English enjoyed the local newspaper, the *Louisville-Lafayette Advance*, which flourished from 1892 to 1897. The *Brooks Vindicator* provided some competition for 1896 and 1897; another newspaper, the *Colorado Sun*, kept going until 1901.

Louisville resident Robert DiGiacomo shows off his 1906 auto.—*Photo, Louisville Public Library*

Starting with the turn of the century, the *Louisville Journal* appeared on the streets until 1912, and the *Black Diamond World*, published by C.W. Brown, was available from 1901 to 1909. Editor Brown, taking note of the invention of the linotype, traveled to Chicago to learn its mysteries. He returned to incorporate the Louisville Publishing Company, which brought out the *Louisville News* until 1915. Louisville's present newspaper, the *Louisville Times*, began publication in 1913.

Lafayette supported newspapers as well. Weeklies included the *Lafayette Inquirer* (1890-1981), the *Lafayette-Louisville Miner*(1892-1894), *Birla* (1894-1895), the *Lafayette News* (1898-1907), the *Lafayette Leader* (1905), the *News-Free Press* (1906-1910), and the *Lafayette Enterprise* (1910-1917).

Caterina Biella starts a new life in Louisville.—*Photo, Louisville Public Library*

The news included any item having to do with coal mining, but also took note of such special events as Mr. Holmes's runaway horse, which demolished his buggy in November 1903. Isabella Summers died in April 1902 after eating pieplant. If Mary Miller went to visit friends in Boulder for the day, that social note appeared in the papers. "Mrs. Fishback, who has been lying at the point of death for the last few days, is somewhat better."[2] "Dr. King left on the morning train for Boulder."[3]

Marriages, births, deaths, the opening of a business, mining accidents—all these bits of news were necessary to the success of a small-town paper. The 1902 papers detailed four explosions at the Sunnyside mine. The Mitchell mine suffered a cave-in. Epidemics were described: Lafayette was laid low from diphtheria in 1903; Louisville suffered from the same disease the following year.

According to the *Black Diamond World*, Louisville adopted a novel way to

Celeste and Angelina Romano on their wedding day.—*Photo, Louisville Public Library*

confine its criminals in November 1903. Since the "old jail was a source of trouble . . . .Anyone could escape from it with very little trouble,"[4] the town council ordered a number of steel cages—they cost $198 each—which they placed in the town hall to house one or two prisoners per cage. Built with one-inch bars placed five inches apart, the device had a little slot at the base of the door through which the prisoner's food was shoved. The cages seemed quite satisfactory, reported the *Black Diamond World.*

The need for volunteer hose teams was essential in towns where almost all of the buildings were made of wood and where formidable high winds regularly swept down from the back reaches of the Rocky Mountains. A single spark often meant disaster. The fear of fire was widespread, and with good reason. On January 24, 1900, Lafayette suffered a disastrous fire when winds from the Rockies blew enough sparks out of woodstoves to cause most of the downtown district to go up in smoke. The hose companies were frustrated because the pressure was too low to carry sufficient water to the blazing buildings. Mary Miller, undaunted by the estimated loss of $100,000 in real estate, calmly announced to the public that Lafayette would be rebuilt as soon as possible.

Six graduate from the eighth grade in 1903: Henry S. Zarini, Ambrose Taylor, George M. Cummings, Jennie Irene Lee, J.E. Schreiber, superintendent, Evelyn Thirlaway, Fred C. Van Arsdale.—*Photo, Louisville Public Library*

The unfortunate Janet Taylor, operator of a livery stable in Louisville, who was killed by her husband in 1902.—*Photo, Louisville Public Library*

Lafayette Bank.—*Photo, Lafayette Public Library*

Most hose companies were composed of thirty men assisted by fifty "fire laddies" on the hook and ladder. (In Boulder, membership in a hose company was a mark of social distinction.) The men practiced and practiced their hose procedure with care; they would parade for any civic occasion or funeral. Lafayette's hose team won the state championship in 1902 and again in 1924.

Fire spread underground as well. On November 20, 1902, flames fanned through the Simpson mine after a series of explosions within a four-day period. The air was so foul below that the firefighters could remain down in the chambers for only thirty minutes at a time. As the lives of the Simpson mules were threatened, the animals were brought up above ground, one by one, by means of the hoist and pulley on the tipple. Most of the mules were blind, for they had spent fourteen years underground in the dark coal rooms. Fire also spread through the Industrial mine near Superior in 1906.

By the turn of the century, those communities that grew "over the mines" began to experience sunken areas in their streets, tilted houses, and sliding buildings. Subsidence, or surface cave-ins due to settling underground over empty coal chambers, began to plague town residents.

The Simpson mine lies under part of Lafayette, while parts of the Acme, Ajax, and Caledonia mines lie under Louisville. The Town of Louisville threatened the Louisville Coal Mine Company with lawsuit; the government charged that the company had pulled down some underground pillars after coal exploitation. However, those pillars left in place also crumbled and fell, while the houses above slid east and west.

In 1912, the Louisville Coal Mining Company was forced to pay up because of subsidence claims and wrote a check to the town for several thousand dollars. Saloon-keeper John Clarkson was not as lucky; his 1902 suit for subsidence damages was thrown out of court. In 1923, the Rocky Mountain Fuel Company deeded some of its land, now called Caledonia Place, to Louisville to quiet claims of subsidence damage.

The two-story Miners Trading Company at Pine and Main streets, the only brick construction in Louisville in the early 1900s, was a victim of subsidence. Even though the damaged building was declared unsafe and condemned by the town trustees in 1909, the structure remained in place for several years, roped off, and provided a popular target for the guards at the Hecla mine in later years who aimed their machine guns at the half-sunken building.

To the north of the coal field, a new "boom" started in 1901: the promise of a big oil strike. Again, geologist Ferdinand V. Hayden was on hand to predict heaven on earth as he had done for the coal industry. Boulder County, he said, "lies over a veritable sea of oil."[5]

Another geologist, B.A. Langridge, further excited the population with his prediction that "the time is not far distant when the eastern edge of the mountains to the Wyoming line will present a forest of oil derricks."[6]

Isaac Canfield was back. The man who opened the Rob Roy mine near Canfield twenty-six years before was now in the oil speculation business. He had been in Ohio and Pennsylvania developing oil fields, had found the first Colorado oil in the Florence field near Canon City, and had returned to the Boulder area an experienced oil driller. And find oil he did. Canfield located an oil fracture on Neil

The consolidated school at Frederick.—*Photo, Colorado Historical Society*

The school at Dacono.—*Photo, Colorado Historical Society*

McKenzie's land near Niwot, a few miles from the Northern Coal Field.

Louisville had its oil "boomlet" as well. The *Denver Times* reported on July 17, 1902, that the Louisville Oil and Investment Company had found oil on the James ranch south of town. "The managers of the well are confident they have a producer, and it will be a short time before something definite can be learned about the value of the strike."[7]

Alas, the highly touted oil boom in Boulder County came and went rather quickly. There was, indeed, oil below ground, but the nature of the fractures made it impossibly expensive to extract for a highly competitive market.

With the development of the Baum coal mine in Weld County in 1906 came the establishment of three new communities along the tracks of the Union Pacific Railroad. Known as the Carbon Valley, the somewhat wooded area was characterized as "infested with rattlesnakes, coyotes, cotton tails and jack rabbits"[8] as well as prairie dogs and mean mosquitos.

Dacono was settled first in 1901, but was not incorporated until 1908, when it numbered 275 residents. Charles Lockard Baum may have christened it "Dacono" after three women—his wife Daisy and her two best friends, Cora Van Voorhies and Nora Brooks.

To the north, the town of Frederick was incorporated in 1907, named after Frederick A. Clark, who owned much of the land thereabout. In four years' time, Frederick grew to a population of four hundred residents whose backgrounds represented thirty–five nationalities. Local citizens read the *Frederick Press*, a weekly journal that lasted one year, from 1908 to 1909. Another weekly, the *Carbon Star*, entertained readers from 1915 to 1917. Later on, the *Coal Ridge News* was Frederick's newspaper from 1921 to 1925. The *Farmer and Miner* began publication in October 1932.

Jacob Firestone founded another little town in the Carbon Valley, which was north of Frederick, in 1908 and named it after himself. A number of mines owned by Denslow Coal and Land Company, which included the Davis and Grant mines, supported the new town's one hundred residents. For a time, Firestone had a good water well; each family was allowed two buckets per day from its reserves. The town's main street took on an old-country look as housewives walked home with buckets of water balanced on their heads. The well was also used by some Frederick residents. The better-quality drinking water was sold in the area for fifty cents per barrel. On wash day, housewives used the hard water that was pumped up from the mines nearby.

Residents of the Carbon Valley did not have to suffer with company stores as such. Each little camp had a post office and a school. If a local child required further education, he went off to Greeley to the north. Each of the towns had its saloons; some had second-story living quarters for bachelor miners. Former coal mine inspector John McNeil was trying his luck at mining at the nearby Sterling mine.

East of the three new towns, well into Weld County, lay the town of Evanston, named for founder Dale E. Evans, and site of the Warwick mine, which opened in 1906. Between Evanston and Frederick lay the Slope mine, later called the Frederick, where a large number of immigrants from Greece and Crete settled for a time; they were the focus of labor troubles leading to the "Long Strike" of 1910-1914.

Oil rigs northeast of Boulder, 1902.—*J.B. Sturtevant photo, A.A. Paddock Collection*

Mother Jones in a contemplative mood.—*Photo, Chicago Historical Society*

CHAPTER SEVEN

The coal industry in Colorado was expanding. In 1876, 100,000 tons of coal were mined throughout the state. By 1900, 5,224,364 tons were sent to market. Colorado had become the largest coal producer west of the Mississippi River. More than 11,000 men mined coal in the state; a mere 1,650 miners, or 15 percent, worked in the Northern Coal Field.

After ten years of occasional attempts to unionize Colorado coal miners, a stronger United Mine Workers of America met in Pueblo on October 29, 1900, to reestablish, out in the open, District 15* as a vigorous labor organization.

The U.M.W.A. had been active in the state since 1891, but had not yet signed up a majority of miners in either the northern or southern field. Even so, some 20 percent of the miners in the north were loyal U.M.W.A. men, but they felt forced to keep quiet about their membership. It wasn't long, however, before the union had established twenty locals with 2,470 members in the state.

The northern field was quickly and solidly organized. Most of the miners lived with their families in permanent settlements. A good number spoke English as their native language and did not have to be convinced of the value of union membership. They did not want to move from mine to mine, on the road from Colorado to Wyoming to Utah.

> They had more at stake in the fight to improve conditions at the particular mine at which they were employed, or in the particular town in which they lived. The northern miner was less likely to pack up and move on if wages were low and conditions poor; he stayed and fought for a union.[1]

A committee report from the Colorado General Assembly assessed the situation:

> We further found that a great majority of those employed in the Northern coal fields have, by dint of economy, built and now own their own homes, the towns of Erie, Lafayette, Louisville and Superior being far different from any other coal mining towns which your committee has visited. The towns mentioned are composed of pretty little cottages, kept nicely painted, giving to the place a home-like appearance. Investigation among the members showed that these homes were, most of them,

* District 15 included Colorado, New Mexico, and Utah until 1922.

> built on the installment plan, by making small monthly payments extending over a period of years, the building of such homes having been commenced at a time when conditions in the Northern fields were more prosperous than they are at the present time, and the fact that they did own their own homes, and the further fact that most of the miners in the Northern fields were married and had families, were some of the strong reasons given by the miners in that district why it was not convenient for them to move to some other part of the state when they found that in the Northern fields, with the conditions that now exist, that they were barely able to make a living.[2]

Up to this point, northern miners had been more apt to get what they wanted from the operators and from state officials by conciliatory means rather than direct confrontation. Southern field miners had been following the example of northern miners, but the situation was to change radically in a very few years.

By 1901, relations between the miners and the operators began to heat up. The men wanted the same improvements they had always wanted; their demands sounded like a broken record. No more scrip for wages. (Operator Charles Baum of Dacono used scrip in defiance of the law.) They wanted the Mitchell Day, an eight-hour shift named after U.M.W.A. president John Mitchell. No more company-controlled weigh bosses.

Despite the disparity between what was legal and what was actual, and despite the growing threat to miners' lives by unsafe working conditions, the U.M.W.A. leadership back east was hesitant to call a strike. In January 1901, after a mass meeting in Louisville, a strike was called anyway, without the sanction of the national union board. One-third of the northern miners wanted to walk out after presenting their demands to the operators. They finally did get the U.M.W.A.'s blessing, almost as an afterthought.

A few of the smaller mine owners signed contracts with the men, but the area's largest producer, Northern Coal and Coke Company, which employed some 70 percent of northern field miners, would not cooperate even though the agreement was to last only until the following October. Management did not desire to discuss twice-a-month paydays, the issue of checkweighmen, or the eight-hour day because they did not want to recognize the existence of the union.

For the first time in the northern field, stockades began to appear around some of the mines; a few companies threatened to import Japanese workers or other scab labor from outside the state. By April, however, James Cannon, president of Northern Coal and Coke, the man who had sunk an early shaft on Mary Miller's farm property, conceded to union demands. Wages rose a little—in real money, not scrip—and the miners paid a little less for blasting powder, but the union itself was not recognized as an institution. Even so, the general feeling among labor organizers for the U.M.W.A. was that they had finally made their mark. The "Louisville Agreement" was in force.

By 1903, however, many of the miners were grumbling about losing the gains they had made just two years before. To "help" President Mitchell with the task of further organizing Colorado's coal fields, a diminutive, spectacled, white-haired Irish lady with an acid tongue came west for the union. Mitchell was not totally certain he wanted her assistance, however.

Known to her admirers as the "angel of the miners" and to her detractors as a "vulgar old lady," Mary Harris Jones, or "Mother Jones," as she came to be called, was a hard drinker, a regular user of socially unacceptable language, and had a "perfect genius for agitation."[3] She was described by a Virginia attorney as "the most dangerous woman in America."[4]

At the time of her first Colorado visit in the fall of 1903, Mother Jones claimed to be seventy–four years old, but, as the organizer showed an "unusual ability to grow more than a year older each year,"[5] she may have been younger or older.

Furthermore, Mother Jones was not all that keen on promoting suffrage for women. She violently hated upper class women:

> A lady is the last thing on this earth I want to be. A lady is a parasitical outgrowth of the system we live under. How many wretches in Colorado have lost their lives in the mines to send one society girl to finishing school, give her a trip abroad, marry her off . . . that is, make a lady of her?[6]

After all, she said, Colorado women may have had the right to vote since 1893, but there was still legal slavery at the state's coal mines. "You don't need a vote to raise hell."[7]

After a few public appearances in Colorado, Mother Jones went back to Indianapolis to tell John Mitchell that Colorado was prime, perhaps overdue, for a strike. Although union leadership was cool to her assessment, a strike was called for November 9; U.M.W.A. leaders back east were amazed to learn that eleven thousand Colorado coal miners had walked off the job.

Louisville's Caledonia Boardinghouse for unmarried miners.—*Photo, Colorado Historical Society*

Although she spent most of her energies in the Southern Coal Field, in and out of jail, Mother Jones did visit Louisville on November 21, 1903, to the delight of the miners. The U.M.W.A. officers were not all that thrilled. At this point, the union was considering a proposed agreement with coal management that would be beneficial to the northern field workers but disastrous to southern field men. On her way into Louisville, one of Mitchell's men told Mother Jones to back off, to support the agreement under consideration, to speak against continuation of the strike.

After all, he told her, "you must not block the settlement of the northern miners because national president, John Mitchell, wants it and he pays you."[8] Mother Jones replied with her usual "sweetness":

> If God Almighty wants this strike called off for his benefit and not for the miners, I am going to raise my voice against it. And as to President John paying me . . . he never paid me a penny in his life. It is the hard earned nickels and dimes of the miners that pay me, and it is their interest that I am going to serve.[9]

The agitator titled her remarks before the Louisville crowd, "We Must Stand Together":

> Brothers, you English speaking miners of the northern fields promised your southern brothers, seventy percent of whom do not speak English, that you would support them to the end. Now you are asked to betray them, to make a separate settlement.

J.J. Steinbaugh's store for hardware and mining tools, Spruce and Front streets, 1909. — *Photo, Louisville Public Library*

You have a common enemy and it is your duty to fight to a finish. Are you brave men? Can you fight as well as you can work? I had rather fall fighting than working. If you go back to work here and your brothers fall in the south, you will be responsible for their defeat.

The enemy seeks to conquer by dividing your ranks, by making distinctions between North and South, between American and foreign. You are all miners, fighting a common cause, a common master. The iron heel feels the same to all flesh. Hunger and suffering and the cause of your children bind more closely than a common tongue. . . .I don't know what you will do, but I know very well what I would do if I were in one of your places. I would stand or fall with this question of eight hours for every worker in every mine in Colorado. I would say we will all go to glory together or we will die and go down together. We must stand together; if we don't there will be no victory for any of us.

I know that President Mitchell has sent a telegram to this meeting endorsing a settlement, but John Mitchell is in Boston, we are here in the field. A general cannot give orders unless he is in the field; unless he is at the battleground. Could a general in Washington give orders to an army in Colorado? I know, too, that there are those in your union who would have us do nothing to help our brothers . . . in Cripple Creek. . . .Goodbye, boys. I shall leave a happy woman if I know that you have decided to stand by our suffering brothers in the south. I will see you again, boys, after I have licked the C.F. & I.[10]

Former office of the *Black Diamond World*, 721 Grant Avenue, is used for Union Labor Hospital Association, March 1909.—*Photo, Louisville Public Library*

Daisy Bracegirdle, a young Erie girl, came to Louisville to listen to Mother Jones and, even in the 1980s, was still struck by the power of her remarks. U.M.W.A. District 15 President William Howells reluctantly went along with Mother Jones that evening, and the miners voted 228 to 165 to continue the strike. Mother Jones left on a late train as, George McGovern says, "she seems so often to have been on the point of doing."[11]

One week later, another meeting was held without the benefit of Mother Jones's presence, and the men voted to go back to work on November 30. The strike continued in the southern field. The miners' leaders there felt forced to confront the operators in a more direct manner, a departure from northern concilatory methods. This change of tactics set the stage for more violence to come in a few years. "Union organization had received a staggering blow, the basis had been laid for industrial civil war."[12]

Those benefits won in the north quickly eroded when coal operators realized that the union was weakened throughout Colorado. By 1905, northern miners were back where they started at the turn of the century, or so they thought, even though another eight-hour statute was on the legislative books. The U.M.W.A. virtually disappeared in the southern field; in the north, its membership remained loyal and anxious to organize further.

A shift at the Industrial mine, Superior, around 1905. The miners still use lard-oil lamps on their caps. Two men are holding breast augers. Lower left, very young workers.—*Photo, Louisville Public Library*

By 1906, a young Greek and his cousin had immigrated from the island of Crete, had landed on Ellis Island, and had made their way across the United States to Denver. There Louis Tikas would soon open a coffeehouse catering to his fellow Greek immigrants. He would take out citizenship papers, make application for the job of Denver policeman, and try his hand as a union organizer in the northern field.

Mother Jones was back in Chicago as a charter organizer of the Industrial Workers of the World. She would return to Colorado to give more speeches, take more trains, and spend some time in jail.

A young girl named Josephine Aspinwall Roche, who had started out life as the kind of lady that Mother Jones despised, would move to Denver with her family so that her father could be closer to his growing coal interests in the Rocky Mountain Fuel Company. When Josephine was twelve years old, she asked her father, now treasurer for the firm, to take her down in one of the mines. No, he said, "it would be too dangerous." "Then," said Josephine to her father, "how is it safe for the miners?" The well-born young woman would later study at Vassar College, then at Columbia University. One of her best friends at school would be Frances Perkins, who would later become Secretary of Labor in the administration of Franklin D. Roosevelt. Miss Roche would get a job to which Louis Tikas aspired—an officer with the Denver Police Department. All the while, her training would prepare her for the job of major coal operator in the Northern Coal Field.

The U.M.W.A. union hall is almost finished. Louisville, 1909.—*Photo, Louisville Public Library*

The Louisville hose team on July 4, 1913, laid down 200 feet of hose in 190 yards.—*Photo, Louisville Public Library*

Christmas comes to the Thirlaway meat market, 1910, Louisville.—*Photo, Louisville Public Library*

Sheriff MP Capp with his deputy Al McDonald, 1910.—*Photo, Louisville Public Library*

CHAPTER EIGHT

When thirty-six-year-old U.M.W.A. organizer John Lawson arrived in the northern field in 1907, the union had lost its influence throughout the state. In the south, no union locals operated, at least not in the open. District 15 was in apparent disarray and under suspension by the national office because it was not self-supporting. No threats to strike were contemplated even though working conditions had gravely worsened in both southern and northern fields.

Lawson was attempting to work quietly, even secretly, in the southern field but decided, after little success, that the northern field was the logical place to start again. The existing groups of organized miners in Erie, Lafayette, Louisville, and Marshall would form the base for the U.M.W.A.'s reentry into the Colorado coal mines.

Born in 1871, John Lawson was working in the Pennsylvania coal mines before he was nine years old. His father was a long-time member of the Knights of Labor. The younger Lawson developed his own special way of preaching the union gospel that was powerful indeed. Lawson's tongue was as golden as Mother Jones's was acid; coal mine officials soon found the union organizer's ability to gain the trust of the miners a serious threat to their operations.

During the 1903 strike, Lawson's home was dynamited by Peter Coryell, a disgruntled coal operator in the New Castle district; Lawson's wife and baby daughter narrowly escaped death. Later, Coryell shot and wounded the union man as he walked down a New Castle street. These violent acts against Lawson only strengthened his resolve to work in the union movement and garnered sympathy and compassion for him and his family. In short, both labor and management knew who Lawson was.

George McGovern writes of the rebirth of union organization in Colorado:

> Covert activity shifted to the northern mines, which were relatively unimportant in production and number of miners employed and where the task of union recruitment entailed less risk to life and limb. Here the miners lived under more civilized conditions than those in the Trinidad field. Many owned their own homes in Lafayette, Louisville, and Frederick, sent their children to moderately decent schools, and participated in a democratic community life. They were mostly English-speaking, generally intelligent and industrious men, and although forbidden even the crudest system of redress for discontents at work, their dignity had not been

> warped by the kind of merciless feudalism that trapped the immigrant colonies 200 miles to the south . . . the mining towns of Boulder, Weld, and Jefferson counties had no company-controlled sheriffs or brutalized corps of camp marshals and were in consequence more accessible to union organizers.[1]

Historian Maxine Benson speaks of "medieval feudalism" as well and writes that it had "at least this advantage—that the feudal lord lived with his retainers and usually fought at their head."[2] McGovern may have been too kind in his characterization of the mine guards, as the violence that would take place in a few years would attest.

Unionization of the northern field was relatively easy. Years and years of Knights of Labor groups in place paid off. Lawson moved quietly from meeting to meeting in the area to assess union strength. He had the help of twenty-two-year-old Edward L. Doyle, a newcomer to the northern field who, with his family, had moved to Lafayette from Illinois a year or so before. Within a few years, Doyle would move from local leadership in the Lafayette area to an officer of the U.M.W.A.'s District 15.

Lawson came out in the open in 1908 to call for negotiations with northern operators. Conversations went well for a time. Union demands sounded like the usual miners' broken record played yet again. The men still wanted to be paid twice a month, they still wanted to work an eight-hour day without fear that men willing to work longer would replace them, they still wanted suitable wages for dead work, they still wanted safety improvements underground, and they still wanted the establishment of a grievance committee at each mine. Colorado still led the nation with twice the number of yearly deaths in the mines; the state's secretary of state James B. Pearce called the coal mines "human slaughterhouses."[3]

When northern operators began to pull away from these discussions, Lafayette union president James McCune and secretary Percy Brillhart called for a strike slated for June 19. The operators, sensing a renewed union strength in the area, quickly backed down. On July 14, 1908, seventeen firms operating in the northern field signed a closed-shop contract with the U.M.W.A., the first major union agreement in Colorado.

With the exception of the closed shop, all other items of interest to the coal miners were somewhere on the Colorado statute books, gathering dust, it would seem. With the first closed-shop contract, miners began to realize that they might become a political force in the state. Their solid vote might swing an election. They voted together to support a candidate for governor, Democrat John F. Shafroth. And he won. Perhaps laboring groups were making headway, after all. So it seemed for the next two years.

These events spurred anxious southern field operators into action. They did not like what they saw coming in the northern field, and, apparently, they had a number of "heart-to-heart" talks with northern operators. If the miners could organize, they reasoned, so could the operators.

Management increased its recruitment of immigrants from eastern Europe for scab labor, although, of course, they did not advertise it as such. No longer did they publicly pretend that coal miners were independent, experienced contractors. They would go with the inexperienced, "untainted" foreigner. By 1910, certainly because of the special recruiting efforts of coal mine management, there were more

Strikers at Louisville.—*Photo, Colorado Historical Society*

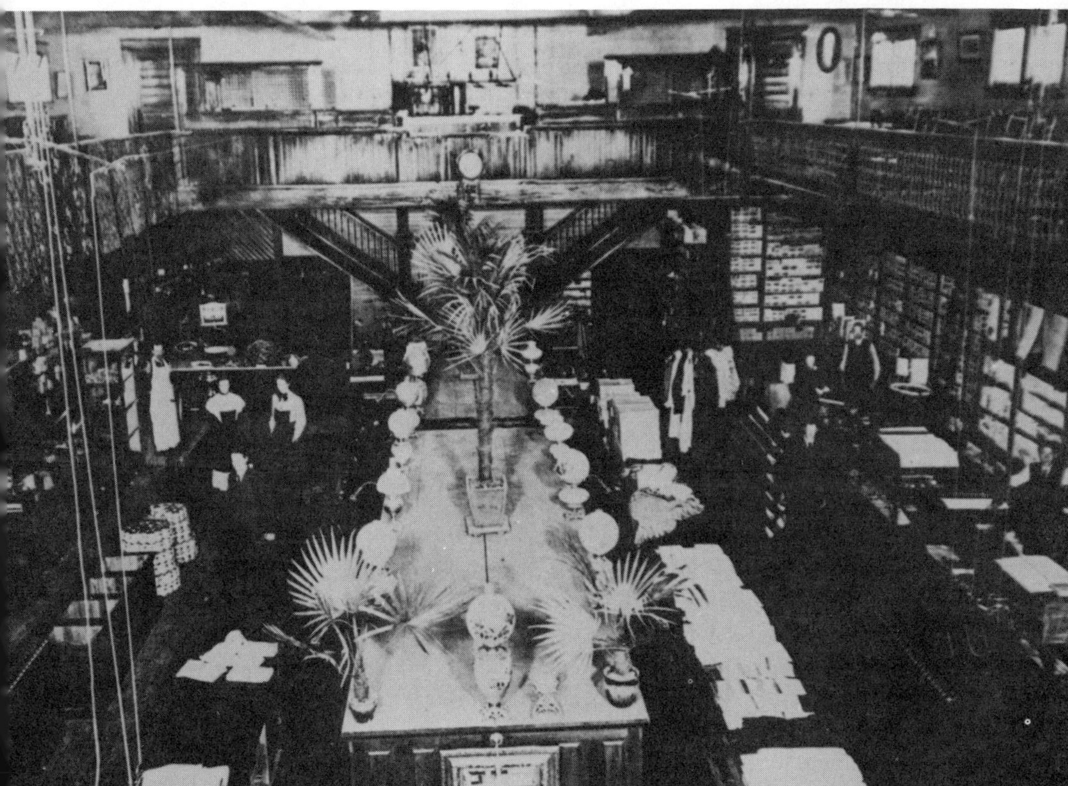

State Mercantile company store, Louisville, 1909, at Main and Spruce streets.—*Photo, Louisville Public Library*

foreign-born workers in the western states than in the East.

Hired guns from the Baldwin-Felts organization based in West Virginia moved into both southern and northern mine properties. Their job was to keep "new employees" on the company land. Searchlights and machine guns appeared near the top of tipples and water tanks. Wooden stockades and barbed-wire barricades now surrounded some of the mines. The owners felt as one and braced themselves to resist any further attempts to unionize the coal industry in Colorado.

Some of the incoming guards initiated or manufactured a fracas here and there so that more of their out-of-work friends could be hired at individual mine sites. This tactic was one of many that gave the Baldwin-Felts employees the reputation of "irresponsible professional guards."[4]

With guards and guns in place, tension in the little coal communities became a way of life. Miners who lived close to coal properties knocked out the searchlights with their shotguns from time to time to end, at least temporarily, the play of light sweeping from cottage to cottage. Bored mine guards enjoyed a little target practice also, aiming their guns toward the miners' homes. Many Louisville residents moved the family sleeping quarters to the basement during those years to avoid a stray bullet in the night. Passengers on the new Denver and Interurban trains crouched low or dropped to the floor of the cars as they passed through Louisville.

By January 1910, local newspapers had stepped up the number of inflammatory articles about miners' demands, the coming shortage of coal, higher fuel prices, and predictions of trouble to come.

At its national convention in Indianapolis that year, the U.M.W.A. voted to seek a pay increase of 15 percent throughout the coal industry in the United States. Upon reflection, however, union leadership decided that such a demand was not realistic; they lowered the demand to 12½ percent.

On March 1, U.M.W.A. leaders presented the wage increase demand to the northern operators. Predictably, several days later, the coal men said no, telling newspaper reporters that coal miners in the northern field were "the best paid coal miners in the world."[5]

In 1910, northern Colorado coal miners averaged $3.10 per day, their wages still determined by the tons mined; there was no pay for dead work. Wyoming miners averaged only $2.65 for the same work. Southern Colorado miners, always at the bottom of the financial scale, brought home an average of $2.48 per day. The Boulder *Daily Camera* warned its readers, "Buy what spring and summer coal you need now as the mines will surely be closed if the miners insist upon the raise."[6]

A few days later, a disorganized U.M.W.A. leadership announced that perhaps a 12½ percent raise in pay was excessive for Colorado miners; now they called for a 5½ percent increase in wages. The operators, surer of their ground, said no again. Toward the end of March, Boulder and Weld county residents knew that the area was to be the stage for either a lockout or a strike. "All contracts between the coal miners who belong to the United Mine Workers of America and the operators of the Northern Colorado lignite fields expired today," reported the *Daily Camera* on March 31.

Lawson and his fellow labor officials saw little hope of any form of conciliation now. The strike was called for the following Monday, April 4. But mine whistles announced the coming strike on Friday, April 1, and many of the men stayed home from work that day. By Monday, 2,700 to 3,000 workers had not reported to the mines. For a time, it was quiet in the empty properties.

Thus started the longest coal strike in Colorado history—four years and eight months. History records a two-year strike in Colorado by coal miners, beginning in 1913. But for the northern field, most union men were out of work for almost five years. During that time, street fights, knifings, and sudden gunfire became commonplace. There were mysterious explosions and unexplained fires. Picket lines and railroad cars filled with frightened scabs were the rule. Children of strikers fought with children of scab workers. At the end, almost five years later, the full-scale industrial war would be discussed by United States citizens throughout the country, would engage the energies of the Sixty-third Congress, and would finally come to the attention of such absentee mine owners as the reclusive businessman at 26 Broadway, New York—John D. Rockefeller, Jr.

During the month of April, the Central Colorado Power Company converted some of its operations from coal to water power to manufacture electricity for its customers. Families of coal miners went on short rations immediately, their little vegetable gardens now a main source of food. Back east, the U.M.W.A. national officers voted to send a weekly three dollars to each striker's family plus fifty cents per week for each child. Bachelor miners on strike received $1.50 each week. U.M.W.A. president John P. White suggested that "Colorado should be a good proving ground for the whole nation."[7] A group of Illinois coal miners voted to send 10 percent of their wages of their wages to their Colorado brothers on strike.

Those who ran the area's company stores were ordered to raise their prices some 40 percent higher than Denver prices for the same food products and clothing. Some local miners were forced to leave their families to look for work in Wyoming and Utah. A number of the immigrants left the northern field for good—Lafayette lost one thousand in population during the strike—and others returned to the area only when times got better.

Mary Doyle, the sister of union official Ed Doyle, opened a restaurant in Lafayette; she charged half-price for meals served to the families of strikers.

More and more barricades appeared around mine properties. Inside the barbed-wire or stockade, now known to the strikers as the "bullpen," Baldwin-Felts guards kept close watch on newly arrived "replacements" from Europe or Virginia or Tennessee, threatening them should they consider leaving the mine property.

The striking miners threatened, too. Somehow, word got through to those would-be scabs coming in on the trains from Denver that if they got off at the Louisville depot, they might be hanged from the nearest trestle. Some jumped the trains to slip around town to head directly to a particular mine. Others thought better of scabbing and went back home to, perhaps, engage in a safer occupation.

Although the telephone had only been in use for a few years, the instrument became an excellent method to send news to striking families without going house to house or boardinghouse to boardinghouse. The telephone operators were the daughters and sons of union men. Daisy Bracegirdle was on night duty at Erie. Eva Lawrence was in charge at the Louisville exchange; May Wiseman was the operator for Lafayette. If trouble broke out at the Acme, or the Hecla, or the Industrial, almost everyone in the northern field knew about it immediately.

George Malcoff had come to the northern field looking for a job. When he sized up the labor problem, he offered himself to the union as a "roamer." Since he was unknown to the operators, he could infiltrate the scab groups within the bullpens and explain to them how they were being used. When the Baldwin-Felts guards weren't looking, some scabs slipped out to help the strikers or left town quickly to look for more congenial employment opportunities. On May 10, ten days after the strike call, a group of eighteen strikebreakers imported to work in the Gorham mine quit in a body.

Jack Green, a Lancastershire coal miner who came to Lafayette in 1906, described the "benefits" given some of the strikebreakers by the management of the Hecla mine. Boardinghouses and apartments, as well as twenty-six cottages, were built to house the new workers. By summer, a "casino" for gambling and other recreations opened inside the bullpen on a hill that came to be called "Hecla Heights." The gambling hall became a great financial asset to the operator who, after paying Hecla workers on Friday afternoon, got the money back over the

Hecla Heights, company housing at the Hecla mine.—*Photo, Lafayette Public Library*

Hecla Heights, casino and water tank, Hecla mine.—*Photo, Lafayette Public Library*

weekend from ill-conceived gambling ventures. Green remembered that, from time to time, a brave Greek or Bulgarian scab would slip away from the bullpen at night to report to the striking men what plans the operators had in mind.

Daisy Bracegirdle remembered that she and her friends would "help" their striking fathers and uncles in an age-old ploy. The sight of pretty young women out for an evening stroll would bring a number of lonely young men to the barbed wire. Later, the bachelors would slip out in the dark to the "other side" and become strikers.

Steve Gorce had immigrated to the northern field from France in 1879 and remembered what it was like for a foreigner to be taken advantage of by the coal operator. He described life at the Puritan mine, then owned by the National Fuel Company, during the strike. A high board fence kept Hungarians, Poles, and other immigrants inside. Even so, the strikers got their message through the stockade, and, one by one, the "new" employees would slip out. All was military at the Puritan, recalled Gorce. Guards at the gates, guards in the company store. To leave the Puritan, permission from the superintendent of the mine was required and seldom given.

As the coal miners' strike went into its second month, rumors were flying throughout the state about non-union miners held captive at mines guarded by the Baldwin-Felts men. Now the guards met the trains and escorted the "fresh labor" to a particular mine. The state's deputy labor commissioner Edwin V. Brake became sufficiently concerned about these rumors that he sent his assistant Eli Gross into the northern field on May 16.

First, Gross visited the Gorham mine near Marshall where he reported that the superintendent attempted to intimidate him and sent an armed guard as escort wherever he went within the mine barricades. Gross sensed he was not viewing the "normal" scope of operations. Next, the investigator looked over the Mitchell mine property. For a time, he was able to stroll away from the superintendent and his guards:

> Upon going to the boardinghouse I found a number of men who said that they desired to leave the camp several times and had been prevented by the guard; that they had been turned back from the exits and told that they would have to get a pass from the superintendent before they would be allowed to leave. Upon application to the superintendent for a pass to leave the camp, the superintendent invariably persuaded them that if they left the camp they would be set upon by thugs on the outside and severely beaten and probably killed. If this talk failed to deter the men from their plan of leaving the camp, he would then say that they could leave the camp provided they left their clothing and properties there. . . .If he believed that they intended to permanently leave the employ of the company, then the pass was preemptorily refused.[8]

He also observed that "the entire atmosphere of these two places was that of men being held at their work at the point of guns."[9]

On June 4, Eli Gross was entering the gates of the Industrial mine at Superior where he was met by a number of miners who claimed they were forced to stay on the job. When Gross questioned the superintendent inside, the company official snorted that the miners' story was nonsense. "However," continued Gross in his report to Commissioner Brake:

> While wandering around the camp I encountered a guard stationed under the tipple who, under a mistaken idea that I was a company official, admitted to me that his orders were to stop of all the men from leaving the camp wherever possible. He said that he had had no trouble with the foreigners, for all he had to do was to order them back to the camp, but that some of these Virginians and Kentuckians and Americans in general were so game that they could not be run back by a display of arms. . . .A number of workmen upon learning that a state official was in camp came to me and signified their desire to leave; all those who desired to leave I took out with me.[10]

Children in the northern field were, of course, affected by their world on strike. On July 10, the *Daily Camera* reported the apprehension of ten-year-old Tony Gabriella of Marshall by local police. Young Tony and a group of his friends had been flinging stones at strikebreakers at the Gorham mine.

One of the methods of control used by Colorado coal operators during troubled times was the securing of services, with or without bribery, of local law enforcement officials. Much to the disgust of the northern operators, Boulder County Sheriff MP Capp could not be bought or influenced. The sheriff—his first name was probably Martin but he never used it—had begun his career in law enforcement with the election of 1908, just as matters were heating up in the coal communities.

Some time in 1910, Capp was approached by a group representing the operators, who asked that the sheriff choose his deputies from a company-approved list of men, i.e., Baldwin-Felts employees. Capp made it clear to both operator and miner that he would not take sides, that his job was to enforce the law. The group hinted that the sheriff might become a little richer if he did what they wanted, but he said no. In fact, MP Capp said later, "The mine operators are the most intemperate men I have ever dealt with."[11]

Certainly one of the reasons Capp felt free to act as his own man was that more residents in the northern field had the right to vote. Further, the children of foreign-born workers attended regular public schools in the area, rather than company schools, learned English quickly, and brought home new cultural ways that helped to move their families into the mainstream of local society:

> The companies did not control the ministers, doctors, lawyers, and storekeepers; the professionals and small businessmen of the northern coal towns were a part of the electorate to whom the miners could appeal for support, and, more important, they could not be easily used by the companies to influence or to coerce the workers.[12]

Sheriff Capp had earned the respect of Colorado Governor Shafroth as well. By August 16, 1910, the governor had met with both operators and law enforcement officials and publicly expressed his confidence in the way MP Capp went about his business. The sheriff gave a statement to the newspapers:

> I have tried to be fair and have tried honestly and faithfully. I know what they [the operators] call fair. They gave me a list of from 75 to 100 men the Tuesday following the declaration of the strike, and asked that I appoint them as deputies.[13]

One month later, an enraged James H. Blood, attorney for the Northern Coal and Coke Company, in the presence of Governor Shafroth, told Capp he would be shot on sight if he entered any Northern Coal and Coke property.[14]

Strikers at Superior with U.M.W.A. placards.—*Photo, Colorado Historical Society*

In jail, to be sure, but dressed for the photographic occasion. Sentenced December 1910 for contempt of court by Judge Greeley Whitford. Standing from left to right: Richard Donald, Ben Hicks, Monroe Hicks, Jess Koenig, George Donald, John Donald, Frank Balek, E.L. Doyle. Sitting, from left to right: Cecil Roose, Duncan Stobs, Robert McBirnie, William Woodhead, William Snow, Joe Fisk, Fred Grayson, George Pansky.—*Photo, Denver Public Library*

None of the coal communities in the Northern Coal Field was immune from violence during the Long Strike. On September 18, Sheriff Capp was called to Superior early in the evening. He and a reporter from the *Daily Camera* hopped on the Denver and Interurban and reached Superior by 9:45 P.M. They found strikers and their families clustered around the depot, which lay five hundred feet from the Industrial mine.

Capp finally learned from a number of witnesses that someone had assaulted a French striker. F.H.L. McCormack, who was both police magistrate and superintendent at the Industrial, and fifteen companions, who had spent the early evening hours at a Superior saloon, were attempting to broaden the fight. Union officials were trying to keep the striking men under check. Both sides were arming themselves with broken whiskey bottles, stones, and the like.

The sheriff stopped the trouble by convincing the strikers to go home and take the matter up with Deputy District Attorney O'Connor the following day. The next night, Mayor William Hake called a special meeting of the town board to determine whether or not Police Magistrate McCormack should be replaced. One of the board members told a *Daily Camera* reporter that he wanted a magistrate who "is not superintending a bunch of scabs, but who will see that we get protection."[15]

In December, after another evening of gunfire—the *Camera* called it a "Christmas celebration"— Mayor Hake was prevailed upon to give up his office.

By this time, the Boulder newspaper editor had looked into his crystal ball and decided not to predict, with any precision, the outcome of the strike. "The strike situation in Northern Colorado fields is unsatisfactory and a crisis considered imminent. A change for the better or worse is likely at any time."[16]

The strike went on and on. Since Sheriff Capp seemed impervious to company "direction," the operators tried another plan of action on November 30. Lawyer James Blood went to Denver District Court and asked Judge Greeley W. Whitford for an injunction against the striking miners. He told the judge that "the officials of Boulder County had not exercised due diligence in preventing atrocities . . . committed by the union miners and their sympathizers in that county.[17]

Judge Whitford issued a restraining order stating that striking miners were forbidden to "interfere with the Northern Coal and Coke Company's operation of its mines," and, most particularly, outsiders were not to threaten those now working for the company. Further, he appointed Baldwin-Felts men as court officials, an action that pleased the operators, both in the north and in the south.

When a big Irishman, John P. "Jack" Cassidy, marshal of Lafayette, tried to stop Barney Fowler, guard superintendent at the Standard mine, from posting Judge Whitford's order, Fowler persisted, and the marshal tried to arrest him. Evidently, more than harsh words ensued. Marshal Cassidy was himself arrested for assault and brought before Judge Whitford on December 17. The court found Jack Cassidy guilty as charged; Judge Whitford told the defendant he would be sentenced sometime after Christmas.

Two days later, three non-union men were beaten up in Lafayette. The victims, with the help of company officials, identified sixteen local men, all of them striking miners, as their assailants. Young Edward L. Doyle, now a Lafayette union official, was named as one of the sixteen. Arrested for assault and congregating, the group came before Judge Whitford on December 23. The judge required that each man sit up front in the jury box for easier identification. The sixteen men told the judge they were aware of the beatings but they did not commit the crimes.

Judge Whitford stated, "The testimony seems to be very conflicting. Somebody has failed to tell the truth."[18] After the judge made a few pointed remarks about "one-sided" government in Lafayette—most officials in that town were union miners—he found the sixteen men guilty of violating his injunction and sentenced each of them to one year in jail.

The judge's decision was front-page news throughout the state. Coal communities in both southern and northern fields expressed outrage. State labor officials were shocked. Governor Shafroth was convinced that "an injustice had been done these men."[19] There was talk of plans to impeach Judge Whitford.

On January 3, during another hearing regarding the activities of Marshal Cassidy, an irate Judge Whitford read one of many threatening letters he had received. His remarks were reprinted in the *Daily Camera* issue of the following day:

> Threats made against the court will not deter this court one iota from pursuing the course it has adopted to restore peace and order in Lafayette. Threats of impeachment, threats of personal violence and threats to kidnap the court are not going to intimidate the court.

Early in 1911, the Colorado General Assembly appointed a committee to investigate the matter, with Representative A.P. Ardourel of Boulder appointed as chair.

On February 7, twelve thousand men and women paraded through Denver to protest Judge Whitford's actions with regard to both Jack Cassidy and the sixteen men. The *Rocky Mountain News* reported, "The meeting was almost a continuous demonstration of enthusiasm."[20] There was talk of a pardon by Governor Shafroth. The matter went to the Colorado Supreme Court.

Hearings at that judicial level brought some unusual information to light. In order to avoid answering questions of a personal as well as a judicial nature, Judge Whitford reviewed his actions, ordered the jailed miners into court, gave them a stern lecture, and released them on February 23. Representative Ardourel's investigating committee also found evidence that the judge may have been bribed and that his personal affairs—was he involved with a woman of ill repute?—were not in good order.

When the miners were brought back from Denver on the Interurban, the little coal towns staged impromptu parties and receptions, first in the streets of Louisville. Then two hundred Louisville miners hopped on the train to accompany the men to Lafayette, where another street party was in progress. A spirited parade, led by the U.M.W.A. local band, escorted the freed men to the union hall. Ed Doyle's sister Mary opened up her restaurant and threw a grand party that night for all who could come. "The jollification continued until a late hour."[21]

When Ed Doyle, a slight but handsome young man, was in jail, a number of families regularly visited the sixteen miners from Lafayette. A young lady accompanied one such family on their jail visit and returned on a number of occasions to visit Doyle. After this unusual courtship, the young lady became Mrs. Edward Doyle in January 1914.

Throughout all the newspaper publicity, Sheriff Capp emerged as an even greater hero in the coal mining towns. The operators, however, retaliated by hiring more Baldwin-Felts guards and detectives for the northern mines.

The high regard for Sheriff Capp probably saved the lives of four Mexican non-union miners who allegedly stabbed to death E.J. "Teddy" Wycherly, a respected Lafayette union official, near a pool hall on Saloon Street on the evening of April 15, 1911. A crowd formed, found the four butcher knives, decided who owned them, and were about to lynch the four Mexicans when Ed Doyle, now president of the Lafayette union local, pushed the assailants into the union hall and pleaded with the crowd for more temperate behavior. Sheriff Capp arrived on the scene in an unmarked car provided by a Boulder newspaper editor, "rescued" the four alleged murderers, and returned to Boulder by a circuitous route through Canfield.

Two days later, the *Rocky Mountain News* described the incident:

> Seeing that Capp, their best friend and hero, who received an almost unanimous vote from the strike district at the last election, faced ruin and failure if he did not save his prisoners, the mob promised to let the Mexicans be taken away.[22]

The General Assembly later passed a resolution praising both Capp and Doyle for "restraining citizens, laboring under intense excitement and anger on account of the unprovoked crime, gathered to avenge the killing of a useful, worthy citizen."[23]

The violence continued. On April 22, E.A. Hockaday, one of Sheriff Capp's deputies, was wounded by a non-union miner in Lafayette and, in returning the fire, the deputy killed the miner.

On June 19, Albert Crabb, a non-union miner at the Simpson, was attacked by

Louisville Brass Band, 1911.—*Photo, Louisville Public Library*

a barrage of eggs. Sheriff Capp arrested seventeen alleged egg-throwers, which included, again, Ed Doyle and Jack Cassidy. This event occasioned a return visit to Judge Whitford's court. The men were fined and jailed, but after another decision by the Colorado Supreme Court in August, all of the men except Doyle and William Crawford (U.M.W.A. secretary for District 15) were released on bond. The two were released a few days later.

Certainly, a major portion of the Louisville community was solidly behind the striking miners. When the U.M.W.A. announced plans to celebrate Independence Day in 1911, the community joined in to form a parade featuring six hundred women and 1,500 schoolchildren as well as 1,200 striking miners. Louisville merchants, as well as a few from Boulder, fed the holiday crowd. Area churches held special prayer services for the strikers. The *Rocky Mountain News* reported that "it was said by the men that this was the first time in the history of Colorado labor troubles where a day had been set aside for praying for the men."[24]

Early in the fall of 1911, the Northern Coal and Coke Company was bought by the Rocky Mountain Fuel Company for five million dollars. Some said it was a quick, easy sale that Northern Coal and Coke did not profit by because of labor and management difficulties. The firm may have lost some two million dollars.

On October 8, a merger between the Mitchell Coal Company and the Economic Coal Company formed the American Fuel Company, whose officers quickly signed a contract with the U.M.W.A. To show their gratitude and to head off threats of economic retaliation from Rocky Mountain Fuel, the U.M.W.A. voted to lend the new management eighty thousand dollars. A state report summed up the situation:

> The majority of the old-time miners in the northern coal fields own their own homes and have built up their towns, and they recognize fully that to allow any combination to ruin a coal mining company which was favorable to union labor would be a blow to them individually as to their property rights, and collectively as to their organization, and they intend to meet any and every move that may be made in this direction.[25]

The miners' money was deposited in the Louisville Bank with disastrous results to that financial institution later on. It would not be the only time that the union loaned money to a coal operator.

With American Fuel in operation, a few union miners could go to work; what mines were open in the rest of the field were using scab labor. But American Fuel's fortunes were not to last; the coal veins played out and the company closed its mines in less than one year.

With each reported incident of violence in the fall of 1911, coal operators pressured Governor Shafroth to bring in the state militia to the northern field. This was another way to diminish the power of Sheriff Capp, who had one more year to go in his term of office.

On November 28, trouble broke out again in Superior. This time it involved the Searcy family, who had come west from Tennessee the year before to take advantage of the strike and work as scabs in the Industrial mine. Dude Searcy, it was whispered in town, had been one of a group to blow up a local farmer's home; moreover, it was generally agreed that Dude—he was christened John Duncan

Harrison Searcy—had a violent temper and a short fuse. In the late afternoon, Dude and his brothers were at the Superior train station waiting for the six o'clock train from Denver. Brother Jim's wife was coming in. A number of striking miners were also waiting for the train to pull in. Words were exchanged.

Dude vowed that he would return to the station later that night to get "the first son-of-a-bitch that would speak to him."[26] Sure enough, at eight o'clock, the Searcys returned to the train station to face some thirty striking miners, including the town's night watchman, Marshal Angelo De Andrea.

Later, no one could prove with certainty which side drew first. After fifteen minutes of gunfire, Dude Searcy was felled from a shot from DeAndrea's gun. It was obvious that Searcy was dying, so the townspeople placed him on the train for Boulder, where he died the next morning at University Hospital.

Meanwhile, telephone lines to the rest of the coal field were humming with rumors and predictions of further violence. Superior, it was said, was about to be attacked by the scabs and the Baldwin-Felts guns. By the morning of November 29, one hundred armed strikers had marched to Superior across the fields from Louisville to face the remaining Searcys and their supporters. Sheriff Capp arrived a short time later with a group of Boulder businessmen. Conferences between the two armed groups went on all day. At dusk, it was decided that the Industrial strikebreakers should retire to the mine's bullpen. Capp convinced the Louisville men to go home. But officials of the Rocky Mountain Fuel Company, the new

Company houses for Superior miners are not exactly homelike.—*Photo, Colorado Historical Society*

owners of the Industrial, apparently did not recognize that Sheriff Capp had managed to keep the peace again. They called for federal troops to be brought in to maintain law and order.

A few days later, rumors were flying again. Someone said that somebody, which side is not known, was planning to blow up the Gorham mine near Marshall. Furthermore, fighting had erupted outside the Hecla and the Matchless mines. Now it was Sheriff Capp's turn to call for the federal troops stationed at Fort Robinson, Nebraska. Colorado's new governor, Elias Ammons, declined Capp's request, saying there was insufficient reason to call in federal men.

When a coroner's jury looked into the death of Dude Searcy, it heard witnesses swear that Dude said to the crowd of strikers the night of his death, "Which one of you bastards will I get first?"[27] Dude had drawn his gun, pointing it directly at Marshal DeAndrea. Some said that Dude fired his weapon first. No matter. The jury of Boulder businessmen found that the marshal was guilty of killing Searcy with felonious intent. Even so Marshal DeAndrea was released from jail on December 8 after a ten-thousand-dollar bond had been posted by thirteen Superior friends. DeAndrea was told to be ready to stand trial in one year's time.

With requests for state and federal military "protection" coming from a number of directions, Colorado's Secretary of State James Pearce and members of his staff decided to visit the northern field on December 8 in order to determine whether or not state militia should be stationed in the area. Pearce studied the matter and concluded that troops were not necessary at this point, saying:

> Certain interests have for so many years been accustomed to break strikes with the militia that it is a difficult thing to break them of the habit. It is much the cheapest and speediest method for them, as the taxpayers of the state pay the bills.[28]

The mines at Marshall are growing into developed coal facilities.—*Photo, Boulder Historical Society, Carnegie Branch Library for Local History*

Officials survey the scene; John Lawson, second from left; Louis Tikas, third from left. Note three-piece suit and gold star.—*Photo, Colorado Historical Society*

CHAPTER NINE

"It was bloody ruthless work."[1]

Leroy Hafen, former Colorado State Historian

All the Greeks in Colorado knew Louis Tikas of Denver. For some time, the immigrant from Rethymnon, Crete, had been the proprietor and part-owner of a Denver coffeehouse at 1746 Market Street, a gathering place for Greek men living in Colorado, coal miners or not. They came together to drink the thick, sweet coffee, to visit—much as they had done in the old country—and to learn the latest developments in the military action in the Balkans.

Some spoke quietly of the hated Leonidas G. Skliris, who had an office across the street, a fellow Greek to be sure, but a man who bought and sold Greek labor to railroad companies and coal operators. Seldom was Skliris in his office on Market Street because the center of his operations was in Utah.

A number of Greek laborers had already returned to Greece or Crete to fight the Turks. But not Louis Tikas. He was different from the others. He wanted to become a real American, and, on April 1, 1910, he had filed first papers for citizenship. He had learned the new language with ease—some said he had received advanced education back in Crete—and was highly regarded for his superior English skills.

A man in his early twenties, Louis Tikas was heavyset but was looked upon as handsome by many who knew him—the ladies included—and his manner was compelling and persuasive. He appeared to possess the qualities of natural leadership, skills that would work against him a few years later.

He was well thought of by Mother Jones, Josephine Roche called him a "lovely person,"[2] and Senator Helen Ring Robinson remembered his courage in difficult situations.

Tikas wanted to be accepted as a new American—he dropped his family name Spantidakis—and did not seem to understand, at least at first, than a number of native Coloradans and northern European immigrants distrusted, even despised, Greeks, Russians, Bulgarians, Serbians, Croatians, Turks and Italians. Many had come to believe that *all* striking coal miners were "ignorant, depraved foreigners, peasants of the lowest and most hopeless class of peoples of Southern Europe."[3] And the Greeks. They were the worst. At best, "murderous rabble rousers."[4]

Sometime during the years that he operated the coffeehouse, Tikas became intrigued with the possibility of becoming a Denver policeman. He applied but was rejected, perhaps because of his alleged membership in the Industrial Workers of the World—regarded by many as the most radical labor group—or perhaps just because he was who he was—a Greek.

Early in 1912, Tikas sold his interest in the coffeehouse to fellow countryman John Kavas. He traveled around the state for a while, perhaps working for the railroad. Tikas was working in Florence when his friend and fellow Greek Steve Fotis sent word from Frederick that Tikas was needed in the northern field. (Fotis came from Crete with the name Stellionos Fotakis, but a railroad boss in Utah "rechristened" him.)

Fotis was working as a scab with a number of Greeks, Italians, and Slovenians at the Slope mine near Frederick. He could see that trouble was coming for him and his fellow strikebreakers. The men wanted to earn more than 39½ cents per ton; moreover, they did not want to continue as scabs. They talked of becoming proud members of the United Mine Workers of America in good standing. Fotis thought his friend Louis Tikas was the man to bring that about.

When Tikas arrived, he and the men met quietly away from the mine to discuss the matter; it was not safe to talk of union membership on the job. According to Fotis, Tikas never worked actively as a coal miner, but the men listened to him, and, on November 18, 1912, Tikas led sixty-three workers on a walkout from the

Mother Jones on parade in Trinidad, 1913.—*Photo, Colorado Historical Society*

Louis Tikas may have fled this union hall in Frederick, 1913.—*Photo, Colorado Historical Society*

Slope mine. Only three miners remained on duty. When the three attempted to set fire to the mine, the others came back temporarily to stop them, fearing that all of them would be blamed for the destruction.

Colorado's union officials took notice of Tikas's leadership abilities, some grudgingly—the Denver Police Department was not the only local institution where prejudice was evident—and put him to work.

Ed Doyle was to say of Tikas later:

> My first connection with him was when he joined the organization. While he had been working as a strikebreaker in the northern field, he came out on strike. Seeing he was an active fellow, we made him one of our interpreters through which to speak to the Greek population and do any translation we needed in Greek.[5]

Three days after the walkout, Louis Tikas may have gotten into more trouble. On the evening of November 21, a number of Baldwin-Felts guards raided the union hall at Frederick. They were looking for a man named Nick Fallas because they had been tipped off that he possessed a gun. They frisked Fallas. No gun. A man named Tikas was alleged to have received the weapon. A number of Greeks were taken to Greeley; they were jailed and subsequently put on trial, but nothing definitive came out of the deliberations and the men were released.

By January 1913, Tikas probably knew a number of the U.M.W.A. officials, including organizers John Lawson and Mother Jones. At some point, the fledgling labor organizer adopted a special sort of "uniform": he almost always wore a three-piece suit with reddish leather puttees (laced-tight leggings) and a large golden star that he pinned to the breast of his suitcoat.

Tikas did not stay long in the northern field. For a time, he worked as an organizer in Lafayette and roomed at the Woolley boardinghouse near the train depot. In his book on Tikas, *Buried Unsung*, Zeese Papanikolas quotes Lafayette miner John Tsanakatsis on Tikas's budding career as an organizer of coal miners:

> But Louis, you know, Tikas, he joined the Union, see, behind with the miners. And he come over to Lafayette, you know. And I tell him, "Now, you better run from here." Because he come over to the boardinghouse. And the guy that opened up said, "Louis, you better run because the guards are after you.". . .He run from the back door. And I was right there at the time.[6]

And run he did, apparently. But not fast enough. They shot him. Later, in a report to the union, Tikas was to state:

> The Louis Tikas upon whom a cowardly assault was made and who was shot by a Baldwin-Felts detective in Lafayette, Colorado, on the thirteenth day of January, 1913, is the same Louis Tikas, the writer of this report to you.[7]

Some said Tikas was offered a lot of money to leave the area. The Boulder *Daily Camera* later reported that a deputy district attorney charged that Louie the Greek, as the papers called him, committed a number of violent crimes in the area and went as far north as Duluth, Minnesota, to blow up a bridge.[8]

When his wounds healed, Tikas departed from Lafayette to continue his union work in the southern field, most particularly at a tent colony at Ludlow.

The strike in the northern field went on and on. More guns. More "incidents." Some out-of-work miners, including Steve Fotis, left their families to find work elsewhere in the West. Now union men did not talk openly about their plans in their own meeting halls. When the U.M.W.A. met in Louisville in February 1913, care was taken to discuss only noncontroversial items on the floor; everyone knew that spies were listening, carrying information back to the coal operator.

At a July meeting of the Louisville town council, the miners convinced the group to pass a resolution condemning the Northern Colorado Light and Power Company for using coal supplies mined by scab labor.

Company housing at the Simpson mine.—*Photo, Lafayette Public Library*

Coal operators continued to be disgusted with Sheriff Capp and his "inability" to conduct himself along company lines. So the mine owners went "federal." After convincing Judge Whitford to withdraw his injunction against strike activity in the northern field, the operators sought new rulings in the federal district court of Judge Robert Lewis.

During the course of deliberations, a number of violent days and nights highlighted activities in the northern field that summer, centered around the Simpson mine near Lafayette. After his investigations, Judge Lewis wondered aloud if the trouble, particularly the "Simpson brawls," were possibly the work of company men rather than instigated by out-of-work miners. In any event, Judge Lewis disallowed any federal injunctions against strike activity, saying:

> The federal authorities should be appealed to only as a last resort. The local authorities should be appealed to first; when they fail, the state authorities, and lastly the power of the nation. It appears that both the local and state authorities claim to have the situation in hand. Therefore, the peace officers must be given credit for the faithful discharge of their duties and under no circumstances, as an independent proposition, will this court take it upon itself to unarm the citizens of that community.[9]

When the U.M.W.A. met in national convention in December, John Lawson was there to report to the leadership that, so far, the union had spent $791,418.02 to sustain the families of the strikers in the Northern Coal Field. With the situation heating up in the southern field, Lawson told the union men that a lot more money would have to be spent in southern Colorado. After discussion, the group finally "bit the bullet" and approved new expenditures.

> Preparation was begun for what was to become one of the most violent, one of the most bitter, and, by all measures, one of the most important strikes in the history of American labor.[10]

Meanwhile, Josephine Roche, who would later figure in strike activity in the northern field, was in Denver preparing to become the city's first "lady cop." The daughter of one of the new owners of the Rocky Mountain Fuel Company, she was now a Vassar College graduate with a Masters of Social Work Degree from Columbia University. She had worked for a few years, both in New York and in Denver, as a probation officer. She was encouraged by Denver friends to apply for the job of inspector of amusements.

Miss Roche successfully completed all of the tests for law enforcement officer and she was hired, with much newspaper publicity. A female law enforcement officer, indeed! The job required that she supervise dance halls and other entertainment spots where illegal activity might occur. She was to deal with gamblers, prostitutes, hustlers, and wayward children, as well as ordinary common criminals. According to her friend Judge Ben Lindsey, Miss Roche could "break up a dance hall row or a riot in front of a saloon better than any experienced policeman."[11]

Her tenure as policewoman was short, however; Denver's city administration changed hands, and the newly elected officials did not appreciate her efforts to clean up downtown Denver. There had been complaints from local businessmen. On April 24, 1913, Miss Roche was handed a discharge notice.

She fought her firing in court, won, and was reinstated. Later that summer, however, she realized it was impossible to work with an administration that was openly hostile to her efforts, and she resigned.

All along, Miss Roche had taken great interest in her father's role as officer with the Rocky Mountain Fuel Company. She followed the strike news, expressing her compassion for the miners and their families. Doubtless her father wondered where his gently raised daughter got these ideas. Vassar? Columbia? Miss Roche later said, "Everything I did prepared me for Rocky Mountain Fuel and for the challenges it presented,"[12] which she would meet in fifteen years' time.

Sheriff MP Capp completed his stormy tenure of office at the end of 1912 with an unsuccessful try for the office of Colorado's secretary of state. No doubt he hoped that the pressures of the last several years would be relieved by more "peaceful" employment. This was not to be for some years. Capp became the warden of the state reformatory at Buena Vista, where conditions were the subject of many a news article. In the midst of controversy over the administration of the facility at Buena Vista, Capp was discharged in 1916. In later years, he operated a pharmacy in Longmont, a quieter occupation indeed, but his life ended violently in 1927, when he was accidentally electrocuted.

On January 14, 1913, Sheriff Sanford D. Buster, a farmer in the area, was sworn in as Boulder County's chief law enforcement officer. Although Buster did not take either side, labor or management, during the rest of the strike, he did not enjoy the full popularity accorded MP Capp. He was just as busy, however. Like his predecessor, Buster investigated each incident of violence in the coal field, went to many an evening meeting, including a few secret ones, and slept with an ear alert for the telephone's ring.

The eight-hour day was back on Colorado's statute books by April 1, 1913. The event was noted by out-of-work miners in the northern field, despite the fact they had no shift to work. They had been on strike for three years by the summer of 1913. It seemed as if the stalemate between miners and operators would go on forever. It was to the south, however, that Colorado eyes turned to read of events that caused the strike to spread statewide and to build in intensity.

On a dusty street in Trinidad one summer Saturday evening, August 16, popular Italian union organizer Gerald Lippiatt, who had spent many months working with northern field miners, was gunned down. Lippiatt had encountered Baldwin-Felts guards George Belcher and Walter Belk on the street. Evidently Belk had dared Lippiatt to a shoot-out, western style, took a shot in his thigh from Lippiatt's gun, and then emptied his firearm with six shots into the labor organizer's body.

Since the U.M.W.A. was to meet in Trinidad on August 18, two days later, a number of northern field miners were already in town. Men from Louisville and Lafayette formed a funeral cortege and journeyed with the body to Colorado Springs, where Lippiatt was buried. At the Trinidad meetings, an empty chair was draped in black to remind the convention of the recent killing. (Three months later, on the same Trinidad street, a crowd suddenly formed around George Belcher as he was walking along. As the crowd thinned again, the Baldwin-Felts guard lay dead, despite the protection of steel breastplates he usually wore.)

Deputy Labor Commissioner Edwin Brake was also in Trinidad at the time of the Lippiatt killing and reported to Governor Ammons that more serious trouble was brewing, that Lippiatt's death had enraged labor men throughout the state. Apparently, Brake's warnings went unheeded.

Ethelbert Stewart of the United States Department of Labor was equally concerned about conditions in the Colorado coal mines and paid a call at 26 Broadway in New York City to see John D. Rockefeller, Jr. He wanted to warn Rockefeller that morale in his Colorado coal mines was such that industrial warfare was a virtual certainty if matters there were not addressed. Rockefeller was not available, Stewart was told, so he went back to Washington, D.C., his message apparently undelivered.

Lafayette's own, Ed Doyle, now the secretary of the U.M.W.A.'s District 15, moved from camp to camp in the southern field; he leased parcels of land near such settlements as Forbes, Delagua, Aguilar, and Ludlow. He and John Lawson ordered a great number of tents as well as other equipment designed for outdoor living. Mother Jones arrived in Trinidad by train to spend her time either giving speeches or living in various jails in the southern field.

Commissioner Brake tried to talk to southern operators; they were not available. He wrote to them. Trouble was about to erupt, he said. Most of the operators did not reply to his letters; the few that did stated that they knew of no trouble between company and employee. Their miners were content, they said.

Brake wrote to the union men as well, attempting to enlist their support to avert

Trouble at the Standard mine, February 1914.—*Photo, Denver Public Library*

trouble. They answered, how could they attend arbitration sessions when most of the matters under dispute were already on the Colorado statute books? Frank Hayes, vice-president of the national U.M.W.A., said to the papers, "I believe conditions in the Colorado coal mines are the worst in America."[13]

Trouble did break out on October 8, when striking miners fought with the strikebreakers employed at the Standard mine near Lafayette. Management brought in one hundred more guards, reinforced by a visit by the state militia.

September 23, 1913, was cold, snowy, and windy. At each coal camp in the southern field, the miners and their families had packed up to move out of their huts and boardinghouses. They streamed down the dirt roads to tent colonies established by union officials. The *Denver Express* described the event:

> No one who did not see that exodus can imagine its pathos. The exodus from Egypt was a triumph, the going forth of a people set free. The exodus of the Boers from the Cape Colony was the trek of a united people seeking freedom.
>
> But this yesterday, that wound its bowed weary way between the coal hills on the one side and the far-stretching prairie on the other, through the rain and the mud, was an exodus of woe, of a people leaving known fears for new terrors, a hopeless people seeking new hope, a people born to suffering going forth to new suffering.[14]

At least 8,000 to 10,000 Colorado coal miners were now on strike. (Barron Beshoar put the number at 12,232.)

Louis Tikas was on hand at the tent colony at Ludlow. He still wore his puttees, his three-piece suit, and the gold star. Now and again he met with labor organizers John Lawson and Mother Jones and had his picture taken with them.

> He always dressed very smart. Wore puttees. But he was well educated. And he was always a gentleman. And old Mother Jones thought there was nobody like him. He called her mother too . . . everybody loved him.[15]

But not everyone liked Tikas, evidently. Some of the labor organizers found the Greek pushy and self-important and had stopped using him as a translator. Tikas was outraged at their attitude toward him and felt he was in the middle of their bickering with one another in their attempts to become more powerful in the organization. He wrote eventually to the International Office of the U.M.W.A.:

> These men, seated around a large table in comfortable chairs in the Trinidad office, smoking their good fifteen cent cigars, riding in automobiles at the expense of the organization and their work limited to the making of a few speeches.[16]

Tikas went on to say that some of the union men appeared to be jealous of his leadership abilities, that they feared that the Greek wanted their "soft lap"[17] jobs.

Since Tikas appeared to be growing in leadership, with or without the support of the U.M.W.A., and since the women reportedly liked the young Greek, he came to the attention of the Baldwin-Felts guards and the militia in the southern field, as he had in the north, particularly to a lieutenant named Karl E. "Monte" Linderfelt, a professional mercenary who had fought both in China and the Philippines. On one occasion, Lieutenant Linderfelt is alleged to have said publicly to the striking miners, "I am Jesus Christ, and my men on horses are Jesus Christs, and we must be obeyed."[18] On December 30, 1913, Linderfelt allegedly had Louis Tikas beaten up by members of the state militia.

Karl Linderfelt of the state militia.—*Photo, Colorado Historical Society*

Members of the state militia pose in front of the Ludlow Saloon.—*Photo, Colorado Historical Society*

As winter came on, the families who moved into the tents became accustomed to the life and did the best they could dealing with substandard conditions of sanitation and the like. After all, the conditions were not much different in the camps on the mining properties they had just left. The presence of these tent colonies was irritating to some, including Governor Ammons, who felt that they were somehow unlawful. He said, "the tent colonies were, of course, silent pickets."[19]

Nineteen thirteen was one of the hardest winters on record in Colorado history. On December 1, after a snowfall lasting three days and nights, most areas received more than forty-five inches of snow. All traffic, including that along the railroads, stopped for a time. The management at the Williams Hotel in Louisville had fifty unexpected guests, refugees from the winter storm. Lafayette resident Elmo Lewis remembers that his father had to tunnel from the house to the chicken coop and to the barn so that the stock might be fed.

During the first three months of 1914, reports of violence in the northern and southern fields seemed commonplace. A mine guard killed, the young son of a miner beaten, fistfights between striker and nonstriker, a girl kicked in the breast by a militiaman on horseback—newspapers published accounts of these events each with their particular editorial bias.

Sunday, April 19, was Greek Easter at the Ludlow camp. Everyone was scrubbed and dressed in national costumes for the special holiday. Louis Tikas

After three days of snow, the scene in downtown Lafayette.—*Photo, Lafayette Public Library*

dressed in festive clothes he had brought from Crete. The women pooled their meager food supplies to prepare a memorable feast with dishes from a number of cultures.

Those enjoying a game of baseball near the tents were startled by five militiamen on horseback who rode through the diamond, dispersing the players. Later, someone was to testify that one of the men called to the Ludlow residents, "Have your fun today; we'll have our roast tomorrow."[20]

At ten o'clock the following morning, April 20, after three short explosions set off by the militia, shots were exchanged between the Ludlow strikers and the state men. No one can say with certainty who fired the first guns. The shooting continued steadily throughout the day and night. Although some of the women and children were evacuated to a nearby arroyo, a number were forced to climb down into earthen pits dug below some of the tents.

Shortly before 7:00 P.M., Lieutenant Linderfelt and his men charged the colony; in the dusk, they torched some of the tent homes with coal oil. For some reason, Louis Tikas and James Fyler, perhaps two other strikers named Rubino and Bartolotti, moved toward the militia lines. Were they called there on a ruse? Were they coming forward to seek a cease-fire?

According to an alleged witness of the affair, electrical engineer Godfrey Irwin said he was standing on a hill above the Ludlow camp and watched as Lieutenant Linderfelt approached the unarmed Greek, still dressed in his Easter finery, and bashed him over the head with a Springfield rifle. Irwin told his story to the New York *World*, which printed the following on May 5: "Tikas fell face forward. As he lay there, the militiamen fell back. Then they armed their rifles and deliberately fired them into the unconscious man's body." Linderfelt is alleged to have shouted to his men, "That damned son-of-a-bitch! He tried to escape!"

All of the men who had come with Tikas—Fyler, Rubino, Bartolotti—were shot to death. Later, some bitter miners said the dead men were strung up to a telephone pole and were used for target practice. However, when the body of Tikas was moved to the union hall at Trinidad several days later, the mourners saw just three gunshot wounds in his back.

The evening of the battle at Ludlow, an outraged Ed Doyle sent the following telegram to President Woodrow Wilson:

> Striking miners and families shot and burned to death at Ludlow. Mine guards with machine guns riddled tents of striking miners and set fire to the colony. Four men, three women, and eleven children murdered. State not only fails to protect, but allows uniforms and ammunition of the commonwealth to be used to destroy the lives of workers and their families. We shall be compelled to call on volunteers in the name of humanity to defend these helpless people unless something is done. The colony is burned to the ground.[21]

This telegram was reprinted in the *Denver Post* on Tuesday, April 21.

Another telegram was sent to John D. Rockefeller, Jr., by his Colorado manager that described the disaster differently:

> Following withdrawal of troops by order of governor unprovoked attack upon a small force of militia yesterday by 200 strikers. Forced fight resulting in probable loss of 10 or 15 strikers. Only one militia man killed. Ludlow tent colony of strikers

Striking families tend to the children at "White City," the tent camp at Ludlow, 1913. —
Photo, Colorado Historical Society

Red Cross workers study the debris that was once the camp at Ludlow.—*Photo, Colorado Historical Society*

totally destroyed by burning 200 tents; generally followed by explosions, showing ammunition and dynamite stored in them. Expect further fighting today. Militia being reinforced. Suggest your giving this information to friendly papers.[22]

Because the former colony was under tight military control now—no one could come in or out—three railroad employees felt compelled to wire John Lawson with this message: "Bodies of women and children exposed in the tent colony. Please do something with the bodies, for God's sake, for they are human souls and deserving of decent burials."[23]

Since the records listing the inhabitants of the colony at Ludlow were also burned, no proper statement could be made as to the number of deaths at Ludlow on April 20.

William Chenery of the *Rocky Mountain News* wrote such an emotionally charged piece on April 22 that it may have caused his later discharge from the paper:

> The horror of the shambles at Ludlow is overwhelming. Not since the days when pitiless red men wreaked vengeance upon intruding frontiersmen and upon their women and children has this Western country been stained with so foul a deed.
>
> The details of the massacre* are horrible. Mexico offers no barbarity so base as that of the murder of defenseless women and children by the mine guards in soldiers' clothing. Like whitened sepulchres we boast of American civilization with this infamous thing at our very doors. Huerta murdered Madero, but even Huerta did not shoot an innocent little boy seeking water for his mother who lay ill. Villa is a barbarian, but in maddest excess Villa has not turned machine guns on imprisoned women and children. Where is the outlaw so far beyond the pale of human kind as to burn the tent over the heads of nursing mothers and helpless little babies?
>
> Out of this infamy one fact stands clear. Machine guns did the murder. The machine guns were in the hands of the mine guards, most of whom were also members of the militia. It was a private war, with the wealth of the richest man in the world behind the mine guards.
>
> Once and for all time the right to employ armed guards must be taken away from private individuals and corporations. To the state, and the state alone, belongs the right to maintain peace. Anything else is anarchy. Private warfare is the only sort of anarchy the world has ever known, and armed forces employed by private interests have introduced the only private wars of modern times. This practice must be stopped. If the state laws are not strong enough, then the federal government must step in. At any cost, private warfare must be destroyed.
>
> Who are these mine guards to whom is entrusted the sovereign right to massacre? Four of the fraternity were electrocuted recently in New York. They are the gunmen of the great cities, the off-scourings of humanity whom a bitter heritage has made the wastrels of the world. Warped by the wrongs of their own upbringings, they know no justice and they care not for mercy. They are hardly human in intelligence, and not as high on the scale of human kindness as domestic animals.
>
> Yet they are not the guilty ones. The blood of the innocent women and children rests on the hands of those who for the greed of dollars employed such men and bought

* This is the first time the word "massacre" was used in print to describe the events at Ludlow.

such machines of murder. The world has not been hard on these; theirs has been a gentle upbringing. Yet they reck not the human life when pecuniary interests are involved.

The blood of the women and children, burned and shot like rats, cries aloud from the ground. The great state of Colorado has failed them. It has betrayed them. Her militia, which should have been impartial protectors of the peace, have acted as murderous gunmen. The machine guns which played in the darkness upon the homes of humble men and women, whose only crime was an effort to earn an honest living, were bought and paid for by agents of the mine owners. Explosive bullets have been used on children. Does the bloodiest page of the French Revolution approach this in hideousness?

In the name of humanity, in the name of civilization, we have appealed to President Wilson. His ear heard the wail of the innocent, outraged and dying in Mexico. Cannot the president give heed to the sufferings of his own people?

Think, Mr. President, of the captain of the strikers, whose truce with the gunmen was ended with his murder. Think of the fifty-one shots which were passed through the strike leader. Think of his body, which has lain exposed since his infamous killing. Then with that vast power which has been committed to you as the executive of a great nation, attend to the misery wrought by an anarchistic lust for dollars. Without your speedy aid the poor and needy, betrayed by the state, may be slaughtered to the last smiling babe.

The monument at Ludlow erected by the United Mine Workers of America.—*Photo, Colorado Historical Society*

Four deputies, newly sworn in by Sheriff Buster, squint at the camera: Gene Shreaves, Frank Burch, Ivan Crees, and Sam Hamilton, circa 1912-1914.—*Photo, Carnegie Branch Library for Local History*

CHAPTER TEN

On Tuesday, April 21, anxious northern field residents milled through the streets of Erie, Lafayette, Louisville, and Frederick, picking up what they could learn of the horrifying events at Ludlow the day before. Rumor after rumor was examined, reexamined, and passed on. Now the newspapers were coming in from Denver; each told a different story. But most revealed that four striking men at the Ludlow camp had been shot and killed by the militia and that at least two women and eleven children had died by fire or suffocation in one of the earthen pits under a Ludlow tent.

The following day, many had a chance to read William Chenery's emotional editorial as well as other articles, written hastily to meet edition deadlines. That same day, the U.M.W.A. leadership issued a "Call to Arms," which was printed in various newspapers throughout the state and read aloud over the telephone lines reaching the smallest camp in the northern field.

> Organize the men in your community in companies of volunteers to protect the workers of Colorado against the murder and cremation of men, women and children by armed assassins in the employ of the coal corporations, serving under the guise of state militiamen.[1]

In Erie, fifty men quickly formed one such company. Two hundred miners in Lafayette and Louisville volunteered as members of four companies. They oiled their guns and set off for Ludlow and other camps in the southern field. They were certain, as were their families back home, that the war would continue, would spread to the northern field.

At the end of the week, angry members of the Denver Chamber of Commerce called for a boycott of the *Rocky Mountain News* and the *Denver Times*, charging that their articles were inflammatory and based upon hearsay and distorted fact. Editor L.C. Paddock of the Boulder *Daily Camera* was almost speechless with rage at the Denver papers and immediately began to write and lay out an ambitious special section of his newspaper in which he would outline his views as to the recent events in the southern field. He would send this edition to every corner of the state. Luckily, he waited a few days, for he would have more news to comment upon, this time in the northern field.

As more and more was learned of the scene at Ludlow, it appeared that, although the first news accounts tended to be lurid, the reports were apparently true enough.

Sheriff Buster seldom came home to sleep. He moved from town to town and from mine to mine, attempting to soothe and quiet angry and frightened miners. Even so, more guns were being oiled and polished.

Those families who had brought their children's beds up from the basement thought of a stray bullet through the bedroom window and took the beds down below again. Those who could afford it sent family members into Denver or into Boulder to stay with friends or relatives. Boulder travelers on the Denver and Interurban still knelt down on the floor for cover while the train passed through Louisville, particularly near the depot and railroad crossings.

Sheriff Buster worried over the number of his deputies. Were nine men enough to cover those parts of the northern field that lay in Boulder County? He continued to visit groups of coal miners and their families, reassuring them that if they continued to play straight with him, as they always had, he would see that their rights were protected. They listened to him, but they also "heard" the guns at Ludlow and "saw" the glow of tents on fire.

Later, in quieter times, the sheriff stated in a deposition:

> I assured the Representatives of the Union that I wasn't elected for the United Mine Workers, or for the coal operators, but that I was elected for the people, and was standing by them, that I had no axe to grind and their fight was not my fight. . . .After the trouble at Ludlow and the attitude the *Rocky Mountain News* took in the matter, and the officials of the United Mine Workers, I realized that things were in a strained condition in Boulder County. All it needed was a firecracker to go off at the wrong place to start trouble.[2]

On April 23, gunfire was reported at the Gorham mine near Marshall; the sheriff lined up a few deputies and went to investigate. A number of women walking down a deserted street told him they were frightened, they were leaving town for a while, that "scabs" were going to shoot up the camp. "Where are your men?" he asked. They didn't know.

At that point, a French resident walked boldly up to the sheriff and said he would show him where the men were. Moving to the Frenchtown part of Marshall, the man beckoned to the sheriff to enter a seemingly deserted home, gave a whistle that was answered, then opened the door to a cottage filled with waiting men with rifles in position to shoot.

According to the sheriff, they told him, "We are going to protect ourselves and families and not allow anything to happen like happened at Ludlow."[3] Buster talked to the men, assuring them of his protection should anyone start a ruckus and, just before midnight, he convinced the frightened men to return to their homes.

Before Sheriff Buster retired for the night, however, he briefly visited Superior and Louisville to see that all was quiet. He returned to Boulder at 2:15 A.M., only to find out that twelve shots had been fired toward the Gorham from somewhere in the Marshall area. Buster could never find out the truth of the incident.

On his next visit to Marshall, he deputized Steve Gorce, the French immigrant who had helped to "educate" incoming scab workers during the early years of the strike. The sheriff chose Gorce because he realized that the Frenchman "had the confidence of all the Union people."[4] Gorce felt afterward that it was a mistake on his part to accept the commission because his neighbors became suspicious of him and felt he had gone over to the other side.

A few days later, Buster was informed by his Louisville deputy that twenty-five or thirty miners were unaccounted for, couldn't be found. What were they up to? Were they meeting somewhere? Later, the lawman found out the miners had gathered near the Hecla mine for a shootout but were talked out of it by one of the union leaders.

On Monday, April 27, one week after Ludlow, the governor's office informed Sheriff Buster that he was to collect all the machine guns and searchlights installed at the mines owned by the Rocky Mountain Fuel Company—the Mitchell, the Simpson, the Standard, the Hecla, the Industrial, and the Gorham.

At six o'clock that evening, the sheriff attended a large meeting of some four hundred miners in Louisville and told them of Governor Ammons's order. It seemed like a good idea at first, but someone expressed the thought that maybe a mine owner's group would overpower the sheriff and use the guns against the miners. Such statements made the nervous even more nervous. A Rockefeller flag waved at the back of the meeting hall; no miner there, however, needed to be reminded of Ludlow.

Sometime after the meeting, while the sheriff was attempting to calm the union officers, he received a telephone call from a deputy who reported to him there was repeated gunfire at the Vulcan mine, southwest of Lafayette and east of Louisville. The sheriff went outside, and, sure enough, he could hear the shots. The mine whistle at the Vulcan was shrieking.

The Louisville men, standing with Sheriff Buster, suggested that they ring the fire alarm bell in town. No, cautioned the sheriff, that might not be a good idea. Sheriff Buster started for the Hecla mine on foot but came back for a moment and told Pat Powers, president of the Louisville union local, to telephone Ed Doyle and John Lawson in Denver.

As he again moved closer to the Hecla, Buster could hear rifle fire ahead, but he also heard the ringing of the Louisville fire bell behind him. Soon, "the shooting was going on all over town and the men made for their homes, got their rifles, ran out of town and, pointing their guns toward the Hecla, turned loose. We had gotten half way when it was a continuous roar all over town from the shooting."[5]

Someone shouted that the infamous Baldwin-Felts guard, Walter Belk, who had killed Gerald Lippiatt the year before in Trinidad, was inside the Hecla bullpen directing operations. Meanwhile, night telephone operators Eva Lawrence, Louisville, and May Wiseman, Lafayette, were busy at their exchanges. At some point, Eva Lawrence called Daisy Bracegirdle in Erie and, in some agitation, told the Erie girl that the Hecla guards were on the rampage. Thereupon, Eva closed down her switchboard, opened the trap door, and went down the ladder into the basement for safety.

Marshal B. C. Pitchford of Erie thought that Louisville was under attack and that Erie would be next. Did he order that the few electric lights in town be turned off? The sudden darkness must have further terrorized the Erie residents. A group of Erie men gathered in the night to find a wagon to carry them to Louisville.

In his attempts to stop the shooting, the sheriff was fired upon by both sides.

Later, a number of residents charged that the Hecla guards started firing first; others said no, it was the striking miners who got off the first shots. When Buster and his men ran to get out of range of the rifle fire, they heard the machine gun at the Hecla start firing. (Others later maintained that the machine gun was in operation earlier.)

Buster related that "it was about 10:20 P.M. when the shooting commenced and there was a continuous roar from then on, up to about 3:15 or 3:30 in the morning."[6] He got to a telephone and called the Hecla bullpen to order the guards to stop firing into town. Things seemed to quiet down. The men from Erie arrived, jumped off a wagon, and moved in from the east, taking their positions behind a stack of straw. Others were huddled in the dry irrigation ditch near the mine and near farmer Rosencrantz's field.

During the lull, Sheriff Buster decided to investigate conditions inside the Hecla bullpen, but, by the time he got to the stockade, about 4:00 A.M. , the shooting started again, even more fiercely. The lawman ducked inside and ordered the guards to use the machine gun only to keep the miners from rushing the stockade; if they did, he commanded, "Shoot to kill." Farmer Rosencrantz—perhaps it was on order of a Hecla guard—opened the headgate at Hecla Lake. Water rushed out to fill the irrigation ditch, which forced wet miners to flee to the straw stack.

Since there was no telephone service for a time between Louisville and Boulder, rumors grew and were elaborated upon even more. Sheriff Buster was shot. Sheriff Buster was a prisoner of the striking miners. So was County Attorney Harold P. Martin.

The shooting stopped between 12:10 and 12:30 the following noon. Just as well, for it was later determined that more guns from Denver were on the way,

The field near the Hecla mine where the Erie men took cover behind stacks of straw.—*Photo, Boulder Historical Society, Carnegie Branch Library for Local History*

ordered by both sides. Buster and Martin had not been shot. Three men inside the Hecla stockade, however, lay wounded. Pete Steinoff, a Bulgarian strikebreaker, was seriously hurt and lay on the floor of one of the boardinghouses at Hecla Heights. Nick Pappas was wounded in the face; William Bucklin, fire boss at the Rex No. 1, was shot through the hand.

Sheriff Buster called Dr. Wolfer, company doctor for the Hecla mine. Dr. Wolfer declined to make his services available, however, saying he would be killed if he tried to come into the bullpen. Said the sheriff later:

> The condition at the Hecla was one of the saddest sights I have ever seen. The boardinghouse was practically all shot to pieces and had blood all over; and down at the Bulgarian headquarters the poor Bulgarian lying shot and dying. The people were stampeding, wanting to know if I couldn't get them out.[7]

Outside the Hecla bullpen lay wounded men as well. Striking miner Baptist Buffo, an Italian immigrant, was shot in the stomach. Fellow Italian Tommy Ronvino was suffering a jaw wound. And Joseph Sirokman also lay wounded.

Through the long night and morning, the telephone exchanges at Erie and Lafayette were busy relaying information, perhaps some misinformation as well, for the frightened families in the northern field believed that Ludlow had started all over again. Telephone repair crews from Boulder moved cautiously into the

Hecla Heights boardinghouse after the shooting, April 27, 1914.—*Photo, Lafayette Public Library*

The Fort Robinson, Nebraska, men walk their horses along the county road west of the railroad tracks toward Louisville. Note speed limit of 8 m.p.h. on sign right.—*Photo, Louisville Public Library*

A company house after the Hecla shooting.—*Photo, Lafayette Public Library*

northern field; they were armed too. When Eva Lawrence came up from the basement, the Louisville switchboard was working again.

When the Denver and Interurban train came through Louisville that night, not only did the travelers get down on the floor, but the conductor drew down all the train curtains and turned out the lights. A number of Boulder men were deputized to guard the Gorham mine near Marshall and attempted to board a train leaving Boulder. The train did not move, however, for the train crew refused to carry the armed citizens into the coal field. After all, they were union men, too.

On Tuesday morning, April 28, after considerable delay, 128 state militia from Canon City were on the train from Denver, which was moving slowly toward the Louisville depot. Their travel was frustrated by more train crews who would not carry military men into the firing area. Ludlow was fresh on their minds, as well. Another rumor. Walter Belk was not in the Hecla bullpen after all; he was on the train coming toward Louisville. When several miners shot at the approaching train, two parties of militia jumped off the moving cars to chase the strikers, some of whom ran for Coal Creek and submerged for a time. They were rounded up by the militia, however, and all moved toward Louisville on foot, some wet and dripping.

As the state troops settled into Louisville and the surrounding area, the shooting stopped for good, despite rumors published in the Boulder papers of further violence to come. The townspeople were still uneasy; after all, some of the troopers

Federal soldiers pick up their equipment at Louisville station.—*Photo, Louisville Public Library*

had just come from Ludlow or the camps at Forbes or Aguilar. These were the men who had shot at the miners in the southern field.

Within a few days, a group of Louisville women petitioned the town council to close down the saloons for the duration. The drinking establishments did close down at six o'clock one evening, but not before members of the state militia had marched from saloon to saloon, helping themselves to alcoholic refreshment.

When members of the Twelfth United States Cavalry from Fort Robinson, Nebraska, replaced the state militia a few days later—Sunday, May 3—the townspeople were so overjoyed to get rid of the "state police" that they welcomed the federal men with a band concert and a parade through the streets of Louisville. The Nebraska units settled into tents east of town near the baseball field; when cold weather threatened, they moved into boardinghouses and other rooms on Main Street.

As the Fort Robinson men began to gather weapons from both sides, they adjusted to Louisville life and were well accepted by the battle-weary residents. Those families who had fled to Denver or to Boulder early on came back to town. One federal trooper became the town's pitcher for the Louisville baseball team; another stayed on to marry a local girl.[8]

During the last of the troubles in the Northern Coal Field, a "Committee of 500" formed in Boulder with the intention of coming into the coal towns to give residents advice and counsel. Whatever the motives of the businessmen and other Boulder citizens, those who lived in the little coal towns in eastern Boulder County found the Boulder attitude condescending as usual, additional fuel fanning the mining families' dislike of the university town.

On May 10, President Wilson took some of the pressure off by issuing a directive that no more strikebreakers would be imported into the Colorado coal fields. A few strikers and their families lived on in the tent colonies to the south for

Home until the snow flies for Troop F, Twelfth Cavalrymen from Fort Robinson.—*Photo, Louisville Public Library*

a time. One by one, they drifted away to find work elsewhere. The strike continued through the summer and on into the fall; relations between the strikers and operators remained quiet but cautious and grim.

Those state militia at Ludlow were almost immediately subjected to a court-martial held near Golden. Barron Beshoar quotes an unidentified military officer who is alleged to have stated to his men, "We've got to vaccinate them so no court in the land can touch them at some future date. A man's life can't be put in jeopardy twice in the courts and we'll take care of these boys."[9]

All of them were acquitted of wrongdoing, including Lieutenant Karl Linderfelt, the man who bashed Louis Tikas in the head. Relieved, he took off to the Colorado mountains to pan for gold.

Throughout the state, officers and members of the United Mine Workers were under indictment for murder, attempted murder, and conspiracy. The financial coffers of the union were empty; in fact, the union was in debt. In Boulder County, a grand jury investigated the shooting at the Hecla and handed down indictments as well. Some involved state officers of the U.M.W.A. who were alleged to have supplied guns to the striking miners in the northern field. Eleven Erie miners—since they were out in the field behind the straw stack the night of the shooting and thus easily identified—were indicted for the attempted murder of William Bucklin, who was shot in the hand during the fray.

At the end of November 1914, President Wilson called for federal mediation of the longest strike in Colorado labor history. By December 10, it was official. The Long Strike was over, despite the raised voices of some northern field miners who wanted to fight on. The union movement in Colorado was thwarted for a number of years. The right to organize into coal miners' unions, the main issue of the Long Strike, had been lost and would weaken the force of the miners' demands to the coal operators for a long time.

George McGovern wrote from a cooler perspective years after the state's era of industrial warfare:

> The mine guards are all thugs, the militamen trigger-happy rabble, the coal operators heartless profiteers. Yet the more one ponders, the harder it is to delineate heroes and villains. . . .Ignorance and braggadocio affected the strikers to a reckless degree, and the motivations of their leaders were not unfailingly altruistic.[10]

May Day youngsters dance for their parents, 1915.—*Photo, Lafayette Public Library*

CHAPTER ELEVEN

An almost five-year lack of livelihood was not the only loss sustained by residents of the Northern Coal Field. Mary Miller's family bank in Lafayette, overburdened with some ninety thousand dollars in U.M.W.A. loans not repaid, collapsed.

In January 1915, the bank in Louisville followed suit and closed its doors as well. Again, the U.M.W.A. funds on deposit there could not be freed because of a lawsuit involving the American Fuel Company, which had failed during the earlier years of the strike. The Louisville Bank deposits totaled a mere $2,600 when it closed. Many northern field residents lost their life savings. Local businesses were on financially shaky ground as well. Somehow, the bank at Erie, established in 1910, remained stable.

Louisville worked fast—"before Boulder can step in"—to establish the First State Bank of Louisville, which opened on February 8.

Since President Woodrow Wilson had ordered a stop to the recruitment and importation of strikebreakers to Colorado, many of the striking coal miners now had the opportunity to go back to work. The men who had been working in Wyoming and Utah coal mines and elsewhere came home to be with their families and start looking for a job.

Since the world was watching—eastern newspapers and magazines had followed the coal strike and the Ludlow massacre closely—many miners in the southern field gained somewhat better working conditions as well as a raise in wages. Coal miners to the north, however, better off than their southern brothers to start with, lost the many advantages they had worked for since the turn of the century.

John Lawson was on trial for murder. So were Ed Doyle and many other union officers and organizers of the U.M.W.A. At some of the local union meetings, there was talk of a struggle for power going on back east within the U.M.W.A. itself. Lawson and others weren't getting much help from the international office.

During the spring of 1915, John D. Rockefeller, Jr., visited Colorado for the first time. In the southern field, he tramped through the mines, ate meals with the miners, and danced with their wives. He spoke to all who would listen of his Industrial Relations Plan, a program designed to give coal miners the right to air their grievances. It was also designed to thwart formation of a proper labor union in Colorado. In April, Rockefeller traveled north and spent a few days enjoying Boulder society. He dined with the president of the University of Colorado and a number of Boulder businessmen, but he did not venture out into the northern field since he had no financial interests there.

Lawyers were busy throughout the state preparing for the prosecution and defense in the many trials of miners on charges of murder, attempted murder, and conspiracy.

The Paddock family had been in the newspaper business in Boulder since the early 1890s. Early on, editor Lucius C. Paddock regularly expressed his disgust with the conduct of some of the Denver papers. A millionaire, he complained, could come to Colorado to buy up newspapers—he had the *Rocky Mountain News* and the *Denver Times* in mind—with no thought of what he regarded as responsible journalism. Since Ludlow, Paddock had been preparing a special newspaper feature, "Insurrection in Colorado During the Years 1913-1914," which he intended to distribute throughout the state to forty dailies and eighty weeklies. Reports of the national guard's role in the Ludlow affair in the Denver papers enraged him. Paddock had been with the guard at Leadville years before

Louisville U.M.W.A. meeting.—*Photo, Louisville Public Library*

and he supported its members wholeheartedly.

He found the activities of those Colorado women who paraded before state offices in Denver to protest the treatment of coal miners repugnant, offensive, and hysterical in nature. He had been equally upset with those Boulder women whose political efforts to close local saloons had been successful several years before.

Paddock was a Democrat but he did not favor those coal miners or any other class of laborer who desired to join forces in union activity. In the 1890s, he expressed editorial views on the coal strikes in Kansas. After all, he said, both union and non-union men worked at the *Daily Camera's* printing plant and got along very well together. The notion of a closed shop rendered him almost speechless, but not for long.

When the first trial involving the shooting at the Hecla opened in Boulder in the spring of 1915, editor Paddock followed it closely and perhaps attended court sessions personally to report on the progress of the trial. Newspapers did not regularly use bylines at this time, but the articles on the trial seemed flavored—peppered, perhaps—with Paddock's distinctive writing style. The trial involved the alleged attempted murder of William Bucklin, the Rex No. 2 fire boss who had

Athletic beauties, Louisville High School, 1915.—*Photo, Louisville Public Library*

Acme miners atop boxcar with message chalked on side, "We are doing our bit for Uncle Sam."—*Photo, Louisville Public Library*

been shot through the hand. Four Erie men—William Knowles, Arthur Spendlow, Gus Brack, and Dan Griffith—were the only defendants to stand trial. Some of the original eleven Erie men indicted by the grand jury had just "melted away," or had left the area for a time.

Selection of a jury did not go smoothly with both prosecution and defense attorneys challenging for cause the panel of prospective jurors. Defense attorney Mike M. Rinn, a former Boulder County district attorney, challenged the selection of Ed Buster, the sheriff's brother, because of his family relationship. When the twelve were finally chosen, after nine days of challenge and rhetoric, Judge Robert G. Strong ordered that the jury be sequestered at the Hotel Boulderado, the finest hotel and eating establishment in town. No newspapers, said the judge. No mail delivery to the jurors unless carefully screened by officers of the court.

Attorneys for the prosecution included Harold P. Martin and George A. Carlson; the latter's loyalties to the case were somewhat divided, as he was also mounting a campaign for his successful candidacy for governor of Colorado.

Defendant Bill Knowles, president of Erie's U.M.W.A. local, was fifty-four years old; he had immigrated to the United States from England with his family in

1861. He had lost his father in a coal mine accident in the northern field. As were so many residents in the area, Knowles was a farmer as well as a coal miner. He was a family man with three children and had served now and again as mine superintendent, justice of the peace, and member of the Erie Town Board.

Arthur Spendlow was a twenty-four-year-old bachelor who started his mining career when he turned thirteen years old. His father had also been killed in an Erie mine in 1891. At thirty-one, Gus Brack was married with three children. Dan Griffith, thirty years old, was an out-of-work railroad engineer who now lived in Denver but, while a resident of Erie, had served as marshal and deputy sheriff.

The *Daily Camera* somewhat pettishly explained to its readers the reason for delay of the trial on its first day:

> It Took Nine Days to Secure
> A Jury to Try Four Ex-
> Strikers for Blowing Up
> The Hecla Property and
> Juror McPhillips' Bowels
> Get Out of Kilter and
> After One Witness Testified
> Trial is Halted[1]

The embarrassed juror McPhillips felt much better the following day and the trial resumed with the state's second witness. Erie resident Zell Wheeler, who no doubt felt he was between a rock and a hard place, "lost his memory" between the time of the grand jury testimony and the trial. He stated he didn't remember whether or not he hauled any strikers by wagon from Erie to Lafayette on the night of April 27, and, if he had, he didn't remember any of their names. Wheeler was promptly arrested for perjury, later tried, and sentenced to three years in jail.

M.T. Alumbaugh later took the stand to testify that Bill Knowles had asked him first to take a group of strikers from Erie to Lafayette, but, the would-be-chauffeur said, he took sick almost immediately after Knowles's request. Another witness remembered that he had taken a number of the strikers back to Erie after the shooting.

Attorney Tom McHarg, who represented other U.M.W.A. members in subsequent trials, testified that he traveled to the Hecla mine that night and persuaded a few of the men to stop shooting at the bullpen. A number of witnesses described the stack of straw behind which the Erie men were firing at the Hecla. Numerous difficulties with identification ensued. No one seemed able to identify anyone else.

Judge Strong was busy. First, he ordered that Mrs. Dan Griffith could not hold her husband's hand during the trial. Next, he criticized defense attorney Mike Rinn for his loud court delivery. Lower your voice, said the judge.

The court heard witnesses who stated they could hear the Hecla guns at Canfield, some eight miles away. At no point was there a definitive statement proving that one side started the shooting. Some said they heard the Hecla machine gun fire first. Others were just as positive that the miners fired from the outside first. Sheriff Buster testified that he thought the miners started the melee.

During the early part of the trial, a famous lawyer slipped into town and attended a number of strategy sessions with the defense lawyers and their witnesses. Clarence Darrow, who was on his way through town to visit his son in Estes Park, then on to a Montana labor trial where he was the star attraction, interviewed some of the witnesses to prepare them for their courtroom appearance.

Erie telephone operator Daisy Bracegirdle remembers his help with her testimony, after which he said, "She's ready for 'em!"[2] She also recalled that Darrow "gave me a rose."[3] Now and again, Darrow sat quietly in the courtroom and advised the defense. One wonders why Darrow's presence in Boulder escaped editor Paddock's attention. He came again, after the trial, for a special celebration supper.

On April 6, attorney Mike Rinn opened the defense of the four Erie men, telling the court of the night telephone call to Erie from Louisville and its message. The *Camera* recorded Louisville's plea for help in headline form with some artistry:

Come For God's
Sake, Save Us!
We Perish

Macedonian Cry for Help
Was Shouted Across From
Louisville to Erie to
Bill Knowles and the
Rest to Come and
Help People of Lou-
isville Who Mine
Guards Were
Slaughtering[4]

When Eva Lawrence, telephone operator in Louisville, took the stand, the *Camera* immediately dubbed her "Little Eva." She said she had talked to Bill Knowles in Erie that night and told him of the machine gun fire. Since Miss Lawrence's statement was made out of the hearing of the jury, many hours were then spent to determine whether or not she could tell her story to the jury.

Meanwhile, the bored *Camera* reporter complained to his readers that each jury member was receiving five dollars per day for this fiasco as well as enjoying sumptuous meals at the Hotel Boulderado at considerable cost to the taxpayer.

Daisy Bracegirdle then took the stand to describe incoming and outgoing calls on the night of the Hecla shooting. When she was asked by the prosecution if Knowles had talked to John Lawson or Ed Doyle that night, she replied it wasn't her place to listen in on private conversations. The *Camera* took note that Miss Bracegirdle was related to two of the defendants on trial—Arthur Spendlow was her cousin and Gus Brack was her brother-in-law. "She is pretty, wore a red rose on her white dress, and is dead in love with the union boys."[5]

On the twenty-sixth day of the trial, prosecution and defense spent hours in fine oratorical summation, and the case went to the jury before supper. Supper, of course, at the elegant Hotel Boulderado. The next day, April 17, the jury's verdict was headlined in the *Camera*:

Knowles and Griffith, the
Leaders of the Erie
Bunch that Came Into
Boulder County to Help
Raise Hell Convicted
While Their Two Fellow
Defendants are Acquitted After Exhaustive and
Exhausting Trial of the
First Hecla Case Before
Judge Robert Strong[6]

A second Hecla shooting trial for the alleged murder of the Bulgarian strikebreaker Pete Steinoff was slated for 1915 and included as defendants union officials William Hickey, Edward L. Doyle, Joe Potestio, John O'Connor, and Jack Cassidy, the former marshal of Lafayette who felt the wrath of Judge Greeley Whitford in 1910. Tom McHarg was to handle the defense of the men, but somehow the trial got lost in history.

With so many court cases scheduled throughout Colorado, most were not

Patriotic assembly at Louisville, April 6, 1918.—*Photo, Louisville Public Library*

Armistice Day in Louisville, November 11, 1918.—*Photo, Louisville Public Library*

After the war.—*Photo, Louisville Public Library*

followed up vigorously for a number of reasons. The Colorado Supreme Court threw out a number of lower court decisions, saying that the men on trial were in court for punitive reasons; specific guilt for a specific crime must be established.

Leslie Hubbard, a 1916 candidate for the state post of attorney general, campaigned in the coal fields of Colorado and pledged to union leaders that, if elected with their support, he would clear court dockets of the remaining cases against the striking coal miners. Hubbard was elected, and he kept his pledge and saw to it that forty-four cases were thrown out.

Locally, Boulder's district court was devoting itself to a long, protracted court fight between the estate of early Boulder financier A.J. Macky and adopted daughter May Oles. Other items on the court docket were delayed and delayed, only to slip off the court schedule into obscurity.

America's approaching role in a world war was on most minds now. As the sons and nephews of older coal miners prepared to go to war, they speculated on the possibility of "settling the score" with some of the more notorious state militiamen with a little well-placed crossfire along the front lines in France.

At the end of 1916, the now-free John Lawson was elected president of District 15 and Ed Doyle was chosen to serve on the union's international board. U.M.W.A. leaders back east apparently did not approve of these selections and immediately suspended District 15 for failure to be self-supporting.

At the following election, Lawson and Doyle tried again, but this time they were defeated—there were allegations of voting irregularities—and fell from grace with some of the union leadership and, perhaps, some of the working coal miners as well. It was not uncommon, after a debilitating strike of the kind Colorado experienced, for the working miners to reject the leaders who had led them during a disastrous strike period. For a time, Lawson worked with the establishment of the Independent Miners Union—in U.M.W.A. circles, his actions were comparable to an act of treason against the United States—but eventually both he and Doyle left Colorado for a time. Lawson would return to Colorado in ten years to play a different role in the Northern Coal Field.

With the passage of the prohibition amendment in 1919, the northern field offered a new service—the illegal sale of liquor. The city of Boulder had been "dry" since 1909; when its citizens felt the need for alcoholic refreshment, they repaired to "wet" Louisville. Some enterprising miners installed stills in seemingly played-out coal mines and offered their product, "white mule," for sale at four dollars per quart. Sheriffs in both Weld and Boulder counties were kept busy as stills moved from mine to mine.

Nineteen nineteen brought a new nationwide end-of-war strike to both coal fields with 450,000 throughout the nation idle. Since no strikebreakers were on the scene, and no guards with guns, the effort did not cause violence and bloodshed. The strike produced no noticeable improvements in working conditions, either. A major issue was the operators' refusal to pay adequately for dead work, or to pay for dead work at all. The miner was still expected to timber, erect crossbeams, and clear away piles of rock or slag for no or little money. Operators still maintained that coal miners should expect to take care of these matters without pay. Sometimes, the decision to pay for dead work was delegated to the mine foreman; since he had to answer to management, his decision was seldom in the best interests of the individual miners.

Another strike was called in 1921 with even less results. The following year, during yet another strike—unhappily referred to as "the summer of John L.'s vacation"[7]—a group of miners met in Superior on July 6. This time there was the threat of violence when members of the Colorado Rangers, a state law enforcement body formed during the war and commanded by Pat Hamrock of Ludlow fame, broke up the meeting with machine guns.

In 1921, families at Erie had more problems than a nationwide strike to deal with. Coal Creek flooded again, and this time:

> Erie was put almost under four feet of water, the flood poured through the streets like a mill race, carrying away houses, buildings, livestock and every moveable article.
>
> One man reported the loss of 26 head of cattle and several hogs and chickens. Another man had just purchased a brand new piano and moved it into his house. After the flood not a trace of the house or piano could be found anywhere.[8]

On September 6, 1920, Labor Day, sixty-five Louisville residents bought tickets on the Denver and Interurban to travel to Denver to enjoy a widely advertised picnic and baseball game. A holiday spirit was evident as the Louisville families boarded the train. They didn't make it to Denver, however. Near Globeville, another train from Denver, also carrying a holiday crowd on its way to Eldorado

What is left of the Denver and Interurban crash at Globeville, September 6, 1920. Twelve killed, 102 injured.—*Photo, A.A. Paddock Collection, Boulder Historical Society, Carnegie Branch Library for Local History*

"Rushing the growler," 1921, Louisville.—*Photo, Louisville Public Library*

Springs, collided with the Louisville train. Since each of the fifty-five-foot cars weighed 125,000 pounds, the collision was fierce. Thirteen passengers were killed, six of them from Louisville. More than 112 were seriously injured. Upon investigation, confusion over whether or not the train schedules were based upon daylight savings or standard time was a factor in the accident.

With the increasing popularity of the automobile, the Denver and Interurban had been running at a loss for some time. Families of the dead and injured filed suit against the firm, a crushing blow to its already floundering finances. The suburban train, so popular in the northern field, made its last run on December 15, 1926, after eighteen years of service. An inexpensive way to travel from coal community to town was lost to those families who could not afford a car.

One winter evening late in 1922, sixty cars and one float formed a silent parade down Boulder's Pearl Street. Auto licenses were draped over so that no one could identify the members of Colorado's newest organization—the Ku Klux Klan. The white-shrouded figures riding in the cars held up signs reading "Join the Invisible Empire," "100 Per Cent Americanism," and "Watch Us Grow in Boulder."

The Klan movement in Colorado had started the year before in Denver, an outgrowth of the Denver Doers Club. Many Coloradans who had belonged to the American Protective Association in the 1890s were attracted to the Klan and joined up without a second thought—if they were white, male, Protestant, and born in the United States. Those who belonged to fraternal orders, fertile ground for Klan membership, saw the new group as a natural extension of their business and social interests. They were used to organizations with secret passwords and private ceremony.

When the Denver Doers, under the leadership of Dr. John Galen Locke, came out into the open as Klavern No. 1, their influence spread to other parts of the state, including Longmont (Klavern No. 2), Boulder (Klavern No. 3), Lafayette, Colorado Springs, Pueblo, Grand Junction, and Steamboat Springs.

Members of the Klan were against Jews, Catholics, and any citizen or immigrant from a Mediterranean or eastern European country. They felt sorry for

Knights of Pythias sit for portrait.—*Photo, Louisville Public Library*

blacks as an inferior race that needed white protection.

Klan members were against "suggestive dances," "titillating motion pictures," "loose women, roadhouses, and joyriding neckers and petters."[9] Movies were singled out because everyone knows, they said, that Jews in Hollywood made the motion pictures showing Jewish women smoking cigarettes, thus creating an unwholesome desire among "American" women to smoke.

Klansmen wanted to rid Colorado of its civil service system, a haven, they said, for Catholics, Jews, and the foreign born. Furthermore, wine should not be allowed in church services—an obvious insult to Catholics and their religious observances.

In 1924, the Klan organization was sufficiently strong around the state that it was able to "invade" the Republican Party convention that year and take over party machinery, replacing regular Republican candidates with a slate of its own. The Klan sent out specially marked ballots to voters throughout the state, "advising" them to choose candidates for office who would form the "visible government." Klan headquarters in Denver also produced a list of Colorado newspapers unfriendly to Klan interests. These papers should be boycotted. The *Boulder News-Herald* made the list when it printed the following criticism of the Klan:

> The right of free speech would be destroyed. The blacklist and the boycott would be used against all citizens who would not be slaves to the Klan Kaisers. Sheriffs would select Klan juries. Klan judges and juries would dispense injustice under the Klan oath. . . .Fear, hatred, and strife would succeed community cooperation and good will. Heaven would become Hell.[10]

The following year, newly elected Colorado Governor Clarence J. Morley made no secret of his close ties to the Klan and took daily orders from the iron hand of Grand Dragon Dr. Locke, even when Locke later went to jail for failure to pay his income taxes for thirteen years. Both senators from Colorado were Klan supported, as was Denver's new mayor. Klansmen throughout the state were encouraged to run for local office and most won handily. Klan representation in Longmont and Lafayette town councils was evident; Klan membership on Boulder's council was alleged.

In January 1925, the *Rocky Mountain American*, a Klan-supported newspaper, began publication in Boulder. The *American*'s ads, some paid for by frightened businessmen throughout the county, "reminded" its readers to buy products only at Klan-approved firms. An auto lot sold "Klean Klassy Kars." There were "Kash and Karry" groceries, stocking "Klean Kanned" goods. A dry cleaner advertised "Klothes Karefully Kleaned," while a shop sold "Klassy Kut Klothes." A local chiropractor placed an ad in the *American*, as did a shoe repair shop, paint store, and several department stores. *Daily Camera* editor L.C. Paddock earned the Klan's hatred when he regularly referred to the group as the "Komic Kapers Klub."

Chain store operations were just beginning to spread across the nation, and the *Rocky Mountain American* told small store owners that the chains were the work of unscrupulous Jewish business magnates. In an apparent effort to counteract this allegation, J.C. Penney's, a new chain business that started in Wyoming, advertised locally in the *American* of its "Spring Values in our White Goods

Department": Wizard sheets were now available at ninety-eight cents each and Wizard pillowcases could be had for twenty-five cents apiece.

The Klan newspaper further advised restaurants to keep Catholic patrons out by putting up signs stating "Fish served every day . . . except Friday." A number of Boulder restaurants put placards in their windows stating that they would not serve Jews or Catholics.

In April 1925, the *Rocky Mountain American* entertained its readers with this charming piece of poetic invective:

> I would rather be a Klansman
> In a robe of snowy white,
> Than to be a Catholic Priest
> In a robe as black as night;
> For a Klansman is AMERICAN
> And AMERICA is his home,
> But a priest owes his allegiance
> to a Dago Pope in Rome.[11]

The editorial on that date stated that "insane, feebleminded and diseased undesirable aliens are being freely admitted to this country."[12] One wonders about the thoughts of the Polish or Greek coal miner if he had the opportunity to read this newspaper.

James H. Davis, scholar of the period, estimated that there were from three to five hundred Klansmen in the Boulder area.[13] Historian Robert G. Athearn estimated the number as closer to one thousand. Statewide, fifty thousand Coloradans were alleged to be members of the Klan at the height of its popularity.

In the summer of 1925, Boulder's Klavern No. 3 played host to some 2,500 Klan members from Denver, Colorado Springs, and Pueblo, as well as out-of-state visitors. The meeting was led by the Exalted Cyclops, whose platform was lighted by the lamps of some five hundred autos parked in a circle in a field somewhere north of Boulder. When *Rocky Mountain News* reporter Lee Casey took a photograph of the group and hurriedly left the meeting, the crowd was about to

Ku Klux Klan meeting north of Boulder.—*Lee Casey Photo, Denver Public Library*

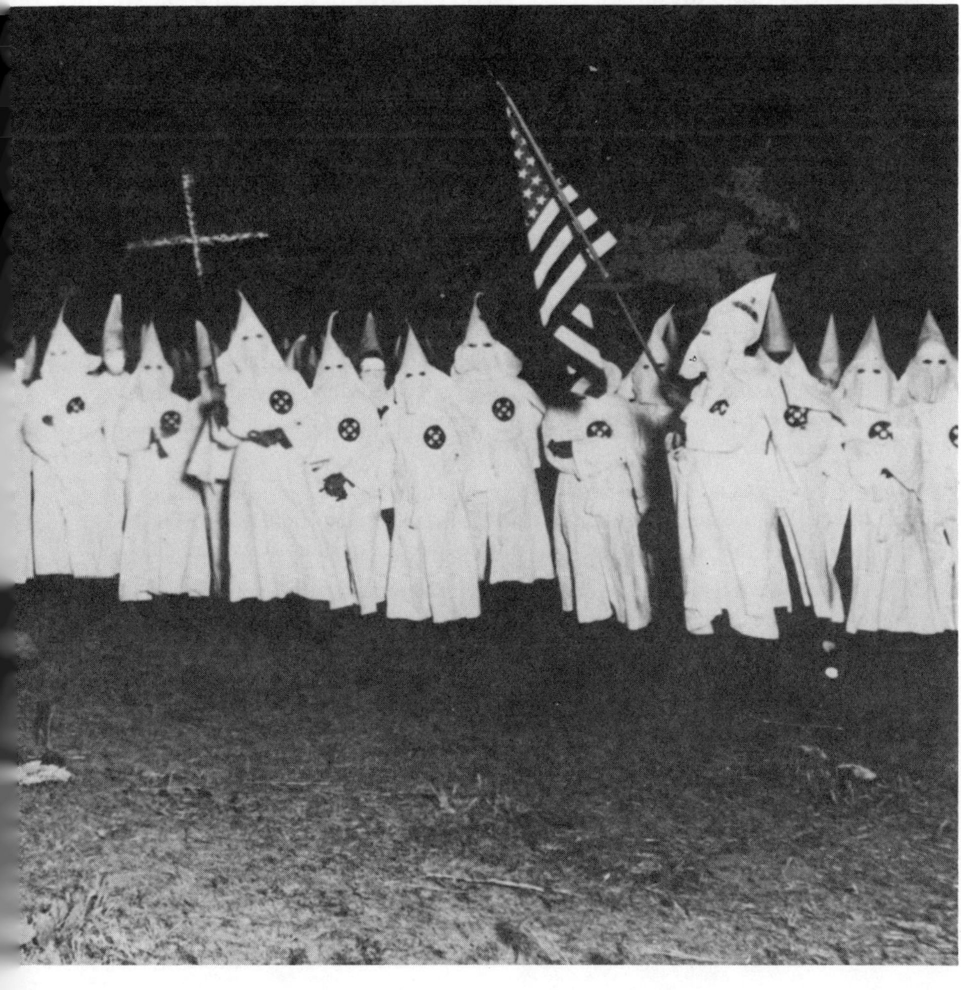

enjoy an old-fashioned barbecue. Klansmen from Lafayette and Louisville were on hand.

On June 9, Longmont was treated to a Klan parade down Main Street. From time to time, members of the Lafayette Klavern would hold a night meeting on a hill east of town. Sometimes they descended the hill to march silently through Louisville or Lafayette, "protected" by a line of white-hooded horsemen. Their "calling cards" were fiery crosses on Catholic lawns left at various locations in the northern field. One night, all the coal towns could see a giant fifty-foot cross aflame on Flagstaff Mountain just west of Boulder.

It was not unusual for white-robed figures to march silently down the center aisle of a church during worship services. On one occasion, they visited the Methodist Episcopal Church in Lafayette where they quietly deposited $103 on the altar. The frightened congregation responded with an additional $57. Lafayette resident Elmo R. Lewis remembered the Klan's visit to the Congregational Church in Lafayette when he was a small boy. He described how scared he was to see the white figures.

Now and then, a ghostly group would "escort" a Greek or Italian to the edge of town; he was then given a clear message not to return. Sometimes during a daylight parade through a coal town, youngsters would run beside the white-robed marchers shouting "Hi, Uncle Harry," or "There's Cousin Ben."

Regarding "the situation" in the northern field in general, the *Rocky Mountain American* had this to say:

> Conditions in the smaller mining towns of this section have become such that they demand state wide attention.
>
> Americans whose fathers and mothers settled Colorado—men who own their own homes, raised their families and have assisted in building our churches and schools are being froze out from some of the mines by a system of wage juggling and discharges to give place to foreigners whose only interest in the United States is to financially gut it and send their money out to bring in the skum of an undesirable underworld, foreign in every way to the good government of a home loving, God fearing white race.
>
> When an American laborer can not vote a clean American ticket or insist that his own country be governed by an American people without having some Jew or Wop foreman "get his job" it is time for the white people of Colorado to take action.
>
> When your house is full of disease-carrying insects and rodents it does not take long for the house wife to get rid of them in the most effective way—this is our America, our home and it is being infected by the morally postulated and diseased of a foreign underworld and it is time to Clean House.[14]

The Klan-supported paper also singled out Louisville as a den of Catholic gamblers. "Wops, every one of them."[15]

The little towns at the northern top of the coal field did not miss out on the attentions of the Ku Klux Klan. Early Dacono resident Verna Veliere remembered that the town's flagpole, located at the center of the business district, was "where the Klan used to burn their flags."[16] She also recalled a classmate in the eighth grade who wrote an essay entitled "What America Means to Me" and was "rewarded" by local Klansmen, who sent her "a great big sheaf of roses."[17]

There were many in the northern field who, while not taking an active role in Klan activities, approved of their actions and took note of the many charities sponsored by the group.

Then, as quickly as it appeared, the Ku Klux Klan, sometimes described as the "refuge of the mediocre man," disappeared from public view. The *Rocky Mountain American* stopped publication on July 23, 1925, just six months after it was established, owing to, as editor Francis delicately put it, "recent difficulties that have arisen in the Klan organization."[18] In its last issue, the paper reported that "Boulder Klavern No. 3 officially died at the stroke of midnight."[19]

The Republicans reclaimed their Colorado party, Democrat William "Billy" Adams took over the governor's office, and, no doubt, many coal miners in the northern field breathed more easily. But the stain of Klan activities in the area was stubborn and hard to remove and affected coal field residents for years after.

For a number of years, organizers from an organization totally different from the United Mine Workers of America had been moving about the southern and northern coal fields. Some of them called themselves the "footloose rebels"; others called them the "Wobblies," or the "I Won't Works." They were members of the Industrial Workers of the World and were about to write a notable chapter in Colorado labor history.

Louisville Brass Band has new uniforms.—*Photo, Louisville Public Library*

CHAPTER TWELVE

> Arise, ye prisoners of starvation!
> Arise, ye wretched of the earth!
> For Justice thunders condemnation,
> A better world's in birth.
>
> No more tradition's chains shall bind us,
> Arise, ye slaves! No more in thrall!
> The earth shall stand on new foundations;
> We have been naught—We shall be All![1]

Nobody liked the Wobblies, apparently. State labor officials feared them, the coal operators couldn't make them out, and what local leadership there was left in the United Mine Workers regarded the Industrial Workers of the World as too radical, too irresponsible, too impractical, too violent, with no real concern for the problems of coal miners. Since 1918, the U.M.W.* had bounced from their membership rolls any miner who was known to carry an I.W.W. card.

Yet this ill-regarded ragtag collection of labor organizers, "bindlestiffs,"† and "footloose rebels" caught the attention of the Colorado coal workers with rousing songs and hearty good humor. The Wobblies supplied the leadership to put on one of the best-run coal strikes in the state's history, "a gem of tactical perfection."[2] Of all the people who were surprised, even shocked, at the success of the Wobblies in Colorado, the most amazed group was the I.W.W. itself.

Wobblie organizers had been in and out of the state since 1905, the year of the group's formation. It was not until September 1925, however, that I.W.W. field organizer Frank Jurich was sent out from Chicago to assess, in a comprehensive manner, the labor situation in the coal mines. Jurich readily determined that the U.M.W. was not an active force in Colorado at this time, so he formed a state I.W.W. unit called Industrial Union 210-220.

Jurich was followed, in a year or so, by I.W.W. general metal and coal

* The United Mine Workers dropped "of America" at about this time.

† or blanketstiffs—those predepression roamers who carried everything they owned wrapped in a small blanket.

organizer Kristen Svanum, who concentrated most of his energies in the southern field.

From the beginning of their presence in Colorado, whispers circulated widely that the Wobblies were agents of "certain" coal operators who paid the "notorious" labor organizers to enter the southern and northern fields to "mix it up," to disrupt matters, to break up what little power the deteriorating U.M.W. had left. In-state labor officials willingly helped to spread these rumors even though there appeared to be no provable basis to the stories.

Another rumor, even more widespread, alleged that there was some financial arrangement between the Wobblies and Texas natural gas interests, whose companies were just beginning to gain an economic position in Colorado. Whatever the source of these stories, no evidence surfaced to characterize them as factual.

Despite various and alleged motives of the Industrial Workers of the World, the organizers found fertile ground in Colorado for a major labor confrontation with management.

The United Mine Workers was in apparent disarray throughout the country. John L. Lewis had been its president since 1920 and was fighting to keep the organization together. Its officers bickered with one another, plotted against one another, and, except for brief organizing forays into the Colorado fields, the union tended to ignore the voices of rising discontent on the part of local coal miners who were increasingly uneasy about worsening working conditions throughout the state.

Although John D. Rockefeller's Employees Representation Plan was not in official effect in the northern field, its philosophy was influential. Northern miners regarded the plan as a company union whose tenets should not spread northward.

As the nation moved toward financial hard times and the Great Depression, those few benefits gained by the coal miner in earlier times were lost, one by one. Most coal companies still did not adequately compensate the men for dead work. The miners still had to timber the coal rooms, clear away rock piles, and lay railroad track on their own time. Although laws were passed regularly at the state level, the men at most mine sites still felt that the company weigh boss was systematically cheating them of their actual daily tonnage.

The Jacksonville agreement, signed in Florida in February 1924 between U.M.W. officials and Central Competitive Field operators—firms located in Illinois, Indiana, and Ohio—called for a minimum of $7.75 for eight hours' work underground per day, as well as fifty-eight cents per ton for narrow work.* Although the Jacksonville scale was in force in Utah and Wyoming for a short time, the pact was soon called a farce by most miners because the majority of the nation's operators went home from the Florida meeting and promptly lowered wages. The first eastern operator to do so was a firm owned by John D. Rockefeller, Jr. By 1925, coal miners' pay had decreased to the 1917 level, a drop of some 20 percent.

The coal industry throughout the United States was in bad shape. A number of newspapers and magazines had these things to say:

---

* Narrow work is the widening of the tunnels underground to accommodate coal cars.

The coal mining industry is poorly organized. It is wasteful and inefficient. It cannot give the American workmen the wages to which it admits they are entitled.[3]

The mining situation in Colorado is similar to that of other soft coal mines in the Mid-West and that of England's discontented miners: it is the natural result of over-production, too many mines and too many miners, inequality of production costs in different mines and localities and the impossibility of enforcing everywhere the same prices.[4]

Coal mining is a seasonable occupation. There are too many coal mines and too many coal miners for the industry in Colorado. This condition has caused over-development and over-production, and, as a result, the average miner does not work over two hundred days a year. With the rate of wages paid in the past, in many cases the earnings have not been sufficient for the maintenance of his family and as a result we have had discontent, strikes, and lockouts.[5]

Miners are suffering from a general over development of the coal industry in the United States. Colorado's population has increased twenty-five per cent and the value of her manufactured products about one hundred per cent in fifteen years, while the coal production dropped from twelve to ten million tons.[6]

Further, the state's Industrial Commission had been in operation for a dozen years and had not been able to gain the respect of coal workers, who perceived that the panel was yet another method used by the coal industry to squash the rights of the coal miner. "You can't convince a single laborer or labor sympathizer in

The Lafayette hose team wins the state championship, July 4, 1924 in Boulder.—*Photo, Lafayette Public Library*

Company houses at the Vulcan mine, Lafayette, October 1927.—*Photo, Colorado Historical Society*

Colorado today that this commission has ever honestly tried to do its full duty; or that it is anything more than a tool of capital."[7] In the late spring of 1927, the commission hired a mine insurance inspector with a name familiar to Colorado coal miners—former U.M.W. organizer John R. Lawson.

Probably the most important factor that led to the success of the Wobblies in Colorado and the strike of 1927 could not be realistically measured—a leftover bitterness from the Long Strike, which had ended badly thirteen years before—unfinished business.

Year by year, the daily wage went down; as was usually the case, northern field miners got slightly higher pay than did those in the southern field, whose wages dropped from $6.20 per day to $5.25. In August 1925, local 2753, U.M.W., was reactivated in Frederick; sixty-seven unhappy miners signed up. By 1927, however, no Colorado coal mine was under contract to the United Mine Workers.

Considering coal miner dissatisfaction with "disappearing" wages, with continuing dangers underground largely ignored by the operators, with the Industrial Commission's lack of effective response, with the threat of the Rockefeller Plan, with the "absence" of the U.M.W., it is hardly surprising that the Industrial Workers of the World established itself in the state with apparent ease.

Card-carrying U.M.W. men became Wobblies "for the duration." In reflecting upon those years, coal miner widow Jessie Irwin James said, "The union men used the Wobblies about as much as the I.W.W. used them, most likely."[8]

The Wobblie men were older; no youthful stalwarts were training to take their places. Fifty-year-old I.W.W. organizer Adam Bell moved into the northern field and established himself in Erie. He worked in the coal mines nearby, familiarizing

Wobblie Adam Bell.—
Photo, Rocky Mountain News

himself with the people and the issues. As Bell moved from coal camp to coal camp, Adolphus Stewart Embree, who was spreading the Wobblie gospel in the Walsenburg area, joined Bell in March 1926 because the I.W.W. leadership readily saw that a solid organization in the northern field was essential to carry along the southern miners.

Embree, a forty-nine-year-old native of Newfoundland, had just gotten out of a Montana penitentiary, having served a term there for "criminal syndicalism." Earlier, he had been a member of the U.M.W. but was expelled for his radical views. The Canadian was well spoken and seemed by his manner to have come from an upper-class background; he was extremely popular with working men wherever he organized. In fact, some said that his I.W.W. colleagues in Chicago sent Embree to Colorado to remove the engaging leader from consideration as a higher official in the organization. The situation in Colorado was perceived as "hopeless . . . his certain failure would not increase his prestige."[9] Embree, in later times, said modestly that this was nonsense.[10]

Another rumor surfaced. "Big Bill" Haywood, one of the founders of the I.W.W., had slipped into Colorado, having returned from the Soviet Union. (Haywood had earlier jumped bail to flee to Russia, an action that cost the I.W.W. some eighty thousand dollars; hence he was under suspension by the labor group.) It was hard to believe that Big Bill would have had the physical strength to return to Colorado, for it was reported that Haywood lay dying in a Moscow hotel from a variety of diseases. (He did die there in 1928.) Another rumor. Haywood had sent in fifty thousand dollars in the late spring of 1927 through Mexican intermediaries to help the Wobblies in Colorado.

Bell and Embree employed a number of techniques of labor organization that seemed foolish at the time, but, foolish or not, they worked. In order to involve an increasing number of dissatisfied miners in the area, the two men formed committees. There were committees of all kinds—the picket committee, the publicity committee, the finance committee, the relief committee, the grievance committee, the defense committee, the strike committee. Everyone had a chance to participate in the effort, an opportunity that had not always been available in the official top-heavy United Mine Workers.

The I.W.W. organizers seemed to have a special ability to deal with those miners who were traditionally at the bottom of the social heap—the recently arrived foreigner, the black, the Mexican, the Oriental. These miners, like everyone else, did not like to be condescended to; they hadn't always gotten a proper reception by the leadership in the U.M.W., whose officers were, for the most part, from Irish, Scottish, Welsh, and English backgrounds.

The Wobblie men apparently dealt with the foreign-born and minority group miners and their families as equals and received in turn a solid loyalty to the I.W.W. Wobblie newspapers and handbills were often printed in several languages, which required the assistance of numerous translators; miners were proud to be thus employed.

What a change from the days when the typical Colorado miner had fair skin, light eyes and hair, and belonged to a Protestant denomination. The following chart illustrates the spread of national backgrounds, thirty-six in number, throughout the coal fields in the state:[11]

American	4,882	Croatian	53
Mexican	3,163	Swedish	49
Italian	1,915	Japanese	49
Austrian	568	Serbian	41
Negro	318	Czech	40
Slavonian	318	Finn	34
Greek	259	Lithuanian	22
English	195	Filipino	14
Bulgarian	178	Belgian	11
Scot	173	Montenegrin	10
German	147	Dane	7
Polish	124	Norseman	4
Welsh	118	Portuguese	2
Spanish	98	Hollander	2
French	87	Swiss	2
Irish	80	Syrian	2
Russian	73	Tyrolean	2
Hungarian	64	Korean	1

Probably the most important organizational technique employed by the I.W.W. was the establishment of auto caravans in the northern field that traveled to camps in the southern area; northern miners were able to show their solidarity with their southern brothers in this way. The cars were loaded with food, clothing, medical supplies, and household necessities that had been donated by Erie, Louisville, and Lafayette residents and merchants.

In October 1926, an auto caravan totaling 150 cars streamed down into the southern field to encourage the miners there, to let them know that a solid front was not only possible but necessary. Officials at John D. Rockefeller, Jr.'s, Colorado Fuel and Iron Company were mystified, even nonplussed, at the arrival

Organizers for the Industrial Workers of the World had a surprise for Colorado in 1927.

of these caravans. How should they deal with these "intrusions" into their coal camps? How should they evaluate the influence these autos filled with boisterous northern field miners might have upon their employees?

By the summer of 1927, the Wobblies felt it was time to move. In the southern field, in Huerfano and Animas counties, I.W.W. leaders quietly agreed that a small walkout would "test the winds blowing out of the coal camps."[12] The publicly stated reason for the proposed three-day walkout, to begin August 8, was to demonstrate concern for the sad fate awaiting Nicola Sacco and Bartolomeo Vanzetti.

Six years before, in Braintree, Massachusetts, the two foreign-born radicals had been tried and convicted of robbery and murder. The two were scheduled for execution. Radical elements in American society believed that Sacco and Vanzetti were tried for their political beliefs and not for their alleged crimes; other Americans of more moderate views tended to agree that the two foreigners were not fairly judged.

On the morning of August 8, many Huerfano and Animas county mines lay idle. Out of a labor force of 1,167 men, 1,132 stayed home. Enough coal miners stayed off the job to astonish the operators and state labor officials, as well as the I.W.W. men themselves.

Since the Wobblies had figured that a work stoppage of three days was necessary to assess their strength in the southern field, they were delighted with their early success and called off the walkout after two days in order, in part, to avoid the wrath of the Industrial Commission.

Whether or not a majority of Colorado coal miners had strong feelings about Sacco and Vanzetti, or even Bill Mooney, is questionable;* what the walkout illustrated to all, however, was that miners were desperately unhappy about their working conditions. The U.M.W. was equally unhappy with the course of these events; District 15 president O.F. Nigro issued regular statements attesting to the U.M.W.'s "painstaking neutrality,"[13] but he did not counsel any union man to cross a picket line.

Now the Wobblies, surer of their organizing base, planned for a strike call in October. At first, they selected October 8 as the day for a state walkout. In order to conform to the Industrial Commission's thirty-day rule, however, they proclaimed they would delay the strike call until October 18. Privately, Wobblie organizers were hustling through the northern field, where, they determined, lay important and influential votes. Meanwhile, the chair of the Industrial Commission, Thomas Annear, announced to the papers and anyone else who would listen to him that the proposed strike was illegal, that the Wobblies did not truly represent the coal miners.

Late August and early September were busy times for the Wobblies, state officials, coal operators, and the miners in both fields. The I.W.W. was securing its northern field base with a series of mass meetings. On September 18, during a large meeting in Lafayette, the men expressed with enthusiasm their support of southern field miners. State officials continued to declare to the press that any strike action by the Wobblie men was illegal. Attorney General William L. Boatwright added his opinion that the I.W.W. was planning an illegal strike. Members of the Industrial Commission said again and again, "This strike is illegal."

While the Wobblie leaders and their followers continued to sing songs and plan auto caravans—some cars came from as far away as California—A.S. Embree and fellow Wobblie Nemmissio Edilla were in Frederick in late September to distribute six issues of the *Colorado Coal Miners' Bulletin*, a crudely mimeographed but widely read series of handbills that found their way into every coal miner's home.

On the evening of October 17, the night before the strike, four mass meetings were held around the state to prepare the men for the following day. The one held in Lafayette was so large—estimates vary from 1,500 to 4,000 in attendance—that it had to be moved to the high school grounds. Embree spoke to the crowd, his podium lit up by the lights of trucks that were parked at the edge of the area. The vote to strike was unanimous. Practically everyone at the meeting was a member of the U.M.W., but the union was not there for them. They put on their "Wobblie hats," perhaps not with pride or enthusiasm, but with determination.

---

* Almost two weeks later, August 22, 1927, Sacco and Vanzetti were put to death.

On the morning of October 18, one-half of Colorado's coal miners stayed home. In the northern field, where 4,500 coal miners were regularly employed, all 4,500 stayed home. (The Columbine mine opened later with a small crew of strikebreakers.)

Throughout all of the activities of the I.W.W. in Colorado during this period, the Wobblie leaders constantly advised the miners not to be the first to use violence. "Violence of any kind will not be tolerated."[14] This from the notorious Wobblies!

Mindful of Ludlow and other periods of warfare between labor and management, the I.W.W. feared that the operators and state law enforcement groups might attempt to precipitate a fight with the miners. A strike handbill stated:

> Hello workers!
>
> You are not to abuse anyone. You are not to strike anyone. Tell them they are hurting themselves as well as the rest of you. There will be no rough stuff on the job. The sheriff's orders to his men are they must not shoot or abuse anyone. If anyone is going to be killed, let it be one of our men first.[15]

On October 26, three days after another huge meeting in Lafayette, a 134-car caravan rolled down the road toward Fremont County in the southern field with more merchant-donated food and clothing. The autos were filled with some five hundred miners and their wives, who waved American flags as they flew down the roadway.

A.S. Embree recalled, "Most of the cars, having been carefully selected, were of the higher priced makes . . . there were only a few Fords but they caused no trouble at all during the entire trip."[16] Gasoline for the cars was donated. Adam Bell and George W. Richardson led the way to Florence in a Star coupe; they were accompanied by Lafayette resident William Henry Lofton, a black man of formidable size and stature. Embree, along with three Bulgarian miners, followed in a Buick. When the caravan passed near Denver, it picked up enough banners to decorate the cars as well as an RKO Pathe′ camera crew, who shot views of the moving caravan for the newsreels.

Embree enjoyed the motor trip. "The route to Florence by way of Colorado Springs took us along some of the most beautiful highways. Weather was at its best, warm and clear."[17]

At least one I.W.W. message was coming through loud and clear, apparently; the following day, the *Denver Post*, sensing the strength of the movement, suggested to Colorado residents that they "put in natural gas and don't be at the mercy of the coal strike."[18]

Now mass meetings in either Lafayette or Louisville were almost an everyday occurence. On November 4, there was a surprising turn of events. Northern field miners voted to arbitrate with the Industrial Commission; two days later, at another meeting, Embree talked the men out of it.

Governor William "Billy" Adams, who had taken office in the spring of 1927 after the tenure of Klansmen Clarence Morley, was uncertain as to the proper stance he should take with regard to the striking miners. Elected on a fairly liberal ticket, Adams was characterized, nevertheless, as full of "economy and conservatism,"[19] a skilled opponent of any "any reform measure that threatens the 'plum tree' of the corporations."[20] The governor was trying to skirt trouble, delaying decisions with regard to the miners' strike, and may well have caused additional problems for Colorado residents. Adams, in line with a campaign promise, dismantled the Colorado Rangers, the wartime state military force formed to protect citizens and their property from the activities of enemy agents, despite the fact that the area had seen no evidence of German sabotage.

The Colorado National Guard had been formed in July 1917—the first such unit in the nation—but Governor Adams was loath to use the force against the coal miners. Instead, those counseling him convinced Adams to set up an additional military force to deal with the situation. He selected Louis N. Scherf, a military aide to Pat Hamrock in the early 1920s—Hamrock was a principal figure in the Ludlow affair years before—to head the new group. Most area newspapers referred to the new force as the "state law enforcement department." No capitals.

Adams did make his views known in late October by issuing the following statement:

> An unfortunate industrial condition in the coal fields of the state by reason of the I.W.W., an un-American organization, having attempted to bring about a strike, and in this attempt has openly and publicly advocated and practiced defiance and violation of the law.
>
> Picketing and intimidation are unlawful. Every man has a right to quit work if so inclined. Every man has an equal right to work and is entitled to protection in that right.
>
> Our state statutes provide that the local peace officers of the counties shall enforce the law, keep and preserve peace. If at any time it shall develop that they are unable to do so, the state stands ready and will give them any and all assistance necessary and use every instrument within the power of the governor to both enforce the law and protect life and property.[21]

Scherf proceeded to drill his twenty men at Walsenburg; he saw that each man had a riot gun, automatic revolver, and cannisters of tear gas. (The riot guns used "dum-dum" ammunition, the first antipersonnel explosive bullets.) Scherf was waiting for a shipment of hard safety hats for the men. He expected the first trouble to come in the southern field and seemed to be looking forward to such an engagement. Now state law enforcement could take to the air. On hand was a Douglas bomber fitted with machine guns and antipersonnel bombs. Three additional airplanes had been ordered from New York.

On November 7, in defiance of the Industrial Commission's ruling against picketing, a large number of strikers and their autos blocked the road between Lafayette and the Columbine mine, the only mine operating in the northern field. Ordinarily, the Columbine, the star property of the Rocky Mountain Fuel Company, employed some three hundred men; now its force was reduced to 150 strikebreakers. Coal had been mined at the Columbine since 1914; the property lay just across the Boulder County line in Weld County.

The Columbine's company town was called Serene, which was not an accurate description of the place, a collection of dirty company houses surrounded by a barricade of barbed wire, illuminated at night by a giant searchlight that was

Serene, site of the Columbine mine.—*Photo, Lafayette Public Library*

Getting ready for Independence Day at Serene.—*Photo, Lafayette Public Library*

Payday at the Columbine.—*Photo, Lafayette Public Library*

installed on the Columbine's tipple.

In addition to living quarters for the employees of the Columbine, Serene boasted a church, a gambling hall, general store, mess hall, a house of prostitution, and the United States post office. At the center of Serene were the Columbine's main shaft, its tipple, a water tank, and a variety of administrative offices.

Because of its location, Serene was "cut off from the world and from outside workers . . . despite its name, dingy little Serene epitomized the conditions the miners were striking against."[22]

On November 8, the miners' cars—some 150 to 200 of them—were on the road again, in stubborn opposition to the regulations of the Industrial Commission. The men had come from another mass meeting in Lafayette, where they had voted unanimously to continue the strike.

Picketing the Columbine became a daily habit. Every morning at dawn, except Sunday, large groups of men, women, and children gathered at the north outer gate of the mine property to sing labor songs and march through the area to the Serene post office, then swing around the mine property and back to the northern entrance. Women and children were at the head of the picket line, the little ones clutching small American flags.

Mrs. Elizabeth Beranek was always there, at the head of the lines, holding a

The tipple at the Columbine mine.—*Photo, Lafayette Public Library*

large American flag aloft as she led the picketers through the north gate. Wife of striking miner Joe Beranek, forty-four-year-old Mrs. Beranek was mother to sixteen children, ten girls and six boys. (She would bear her seventeenth child after the strike.) All of the Beranek sons would eventually work in the coal mines. Because of her "no-nonsense" manner, the Denver papers often referred to Mrs. Beranek as "the Amazon," although she stood a mere five feet, two inches tall.

Sometimes the picketers marched silently through Serene to the beat of a snare drum; more often they sang lustily and long. The old Knights of Labor song had been revised with new words:

Hold the fort, for we are coming,
Union men be strong.
Side by side we battle onward,
Victory will come.

We meet today in freedom's cause
And raise our voices high;
We'll join our hands in union strong
To battle or to die.

The position of the striking men at the back of the lines was a constant source of irritation to the mine guards—there had been two; now there were nine—who shouted through the barbed-wire barricade, "Put your women and children in the rear and come on."[23] "Get to the front of the line, you cowards," they called, their guns trained upon the moving lines of several hundred. "Not on your life," shouted back the men, feeling fairly certain that the guards would not fire on their wives and small children.

One early morning, the marchers saw a new addition to the bullpen, a machine gun mounted upon a truck. Perhaps the sight was chilling, but they all sang together, both the Wobblie songs and the traditional labor tunes, accompanied from time to time by a local brass band.

On November 9, Lafayette learned that union leader Karl Clemens, a Lithuanian immigrant with eleven children, was in jail in the southern field. Louis Scherf was systematically rounding up the Wobblie leaders, jailing them, then moving them from courthouse cell to courthouse cell every two days to thwart the

The Wobblie office at Serene, 1927.—*Photo, Lafayette Public Library*

labor lawyers' attempts to serve writs of habeus corpus. A.S. Embree was caught in Scherf's net as well.

With a variety of Wobblies in jail, the I.W.W. brought out the last in their bag of organizational tricks. In the I.W.W. newspaper, *Industrial Solidarity*, this appeal was featured: "Send all foot-loose rebels to Colorado. This is the hour of glory for every rebel in the economic war on the Colorado Fuel and Iron Company and allied robbers of coal miners."[24]

And they came from all parts of the country. They walked, hitched rides, or rode the rails. Perhaps they sensed that this was one of the last times, perhaps even the last time, the Industrial Workers of the World would have the appeal to pull in a large number of people to help a labor strike.

At the Lafayette mass meeting of November 11, the high point of the evening was the "unmasking" of Sergeant William White, a state law enforcement man, whose orders were to spy on the striking miners. The men hooted and hollered at White; before they threw him out, they shouted, "We are not I.W.W.'s; we are Colorado miners."

The next morning, about one thousand northern miners went on a "field trip"—it was Adam Bell's idea—and visited a number of mines in the area to show the operators that they still "had pep."[25]

The picketing of the Columbine at dawn's first light had become a social event in the northern field. Lafayette residents were often awakened by a badly played trumpet that announced it was time to get up and go to the Columbine. Doughnuts and coffee were often served during the affair; sometimes baskets of food were distributed. Miner Steve Fotis remembered that his wife and children went to picket the Columbine several times; he wouldn't go, he said, because the Wobblies demanded that he not carry a gun.

Eighteen-year-old Jessie Irwin and a number of high school chums decided to join the early morning picketers. "It was to be a lark,"[26] said Jessie. The girls told their parents they were staying overnight at someone else's house and planned their adventure at the Columbine for early morning, November 21.

On November 13, two giant rallies were held in Denver, one at the Grace Community Church with minister A.A. Heist and journalist Frank L. Palmer as speakers. Palmer, the former editor of the *Colorado Labor Advocate*, told the crowd that the picketing "has been peaceful, in spite of the gunmen."[27] Adam Bell spoke at the second rally, held at the civic center. As he spoke, 250 cars steamed in from the northern field on their way to the southern field, the autos festooned with signs, banners, and flags.

The following day, tension picked up outside the Columbine, where the picketers and Weld County law enforcement officials scuffled for a time. The sheriff's deputies held captive one of the pickets; when they attempted to arrest him, five hundred miners "took him back." Perhaps Bell sought to lighten the "seriousness" of the moment, for the group repaired to downtown Boulder to join a forty-seven-auto parade down Pearl Street. After the scuffling incident, however, Weld County Sheriff Ben Robinson ordered that the Columbine's outside gate be closed. No more parading through the mine property and the town of Serene.

On November 16, the early-morning picketers saw that the machine gun that has been trucked in was now installed within the barricade, placed high on the

Lafayette girls plan a trip to the Columbine in 1928: Margaret Gendolo, Jessie Irwin, Rose Tianna, Kathleen Morgan, and Iola Bagdonas.—*Photo, Jessie Irwin James*

water tank. (Later, Erie miner George Hoos said that another machine gun was visible near the top of the tipple.) Barbed wire now surrounded the entire property with growing piles of sandbags evident here and there. The United States post office at Serene was no longer available except to those who lived in company houses at the Columbine. Above Serene, four airplanes regularly patrolled the area.

On November 17, yet another auto caravan of fifty vehicles rolled south to El Paso County to distribute food; again, the northern miners expressed their feelings of solidarity with southern miners.

Changes were occurring in the management of the Columbine's parent company, the Rocky Mountain Fuel Company. Early in 1927, John Roche, first the treasurer and then president of the board, largest stockholder in the firm, died. His forty-two-year-old daughter Josephine inherited his financial interests in the company. In March, Miss Roche attended her first stockholders' meeting with the hope that she might gain control of the board and thus change the firm's policies toward the striking miners. She was only able, however, to choose three of the seven directors on the company board.

Although it is difficult to understand how it came about, somehow Miss Roche

Josephine Roche.—*Photo, Western Historical Collections, University of Colorado at Boulder*

was able to convince a conservative board of directors to hire as vice-president and general manager Merle D. Vincent, a Denver attorney and past president of the Colorado Bar Association, well known for his liberal views. Surely the "old guard" must have strenuously objected to Vincent's published views on labor problems, such as: "Operators have consistently and continuously been at war with labor, with the public, and with each other. . . .When operators in their greed for business cut prices to bedrock, they make their men pass the loss."[28] A statement appropriate to the platform of the Progressive party of which both Vincent and Miss Roche were members, but unpalatable to most operators.

Both Miss Roche and manager Vincent strongly disapproved of the conditions under which Rocky Mountain Fuel employees had to work, but until the heiress gained controlling shares in the firm, she and Vincent had to defer to more conservative attitudes, at least for a time. She did ask for and got the discharge from Columbine employment of guard Walter Belk, the man who killed labor organizer Gerald Lippiatt on a crowded street in Trinidad during the Long Strike.

The number of early-morning pickets at the Columbine was growing. On November 19, it was reported that eight hundred men, women, and children marched around the north end of the Columbine property, singing and shouting.

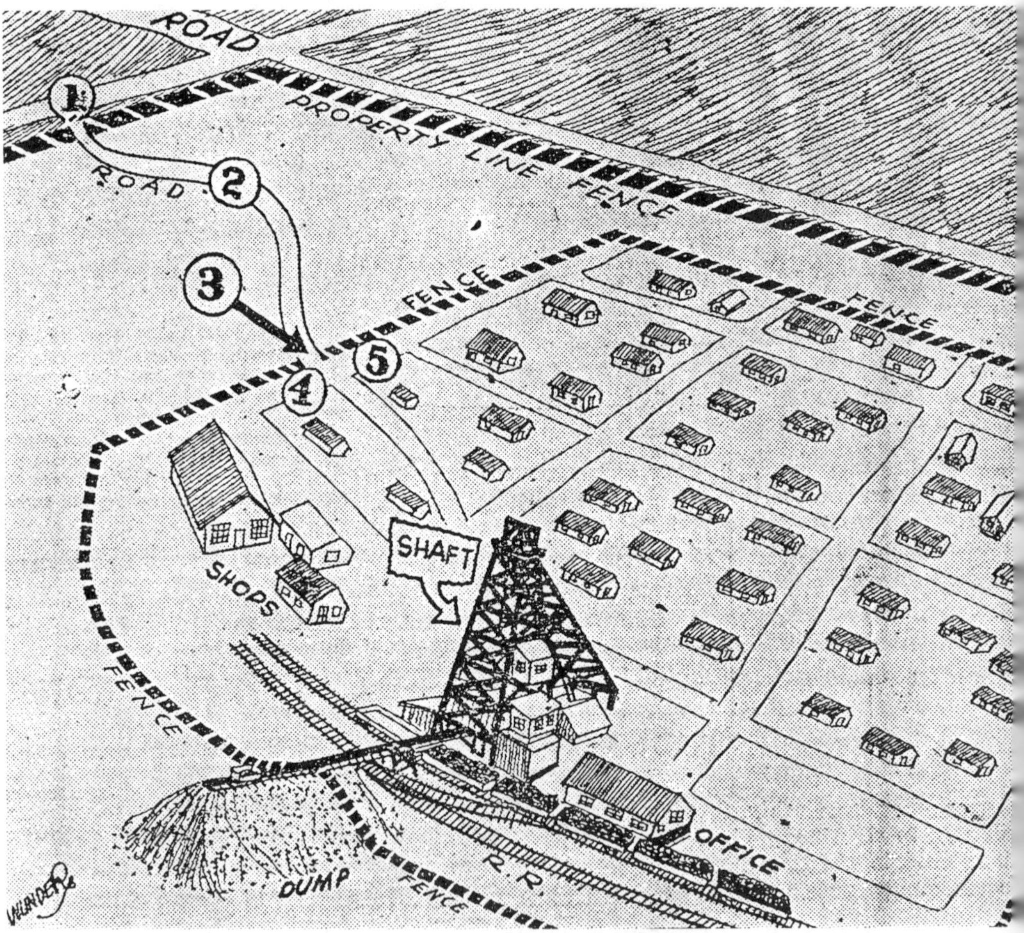

Map of the Columbine mine property. At point No. 1, Adam Bell and Elizabeth Beranek assembled the picketers for their daily parade.—*Diagram,* Rocky Mountain News, *November 21, 1927*

CHAPTER THIRTEEN

Sometime during the late afternoon on Sunday, November 20, Louis Scherf and twenty of his state law enforcement men came up from Walsenburg to the Columbine mine property. His group moved on very short notice; Scherf was concerned that his force did not yet have steel helmets. He sent to Denver for them, and they were delivered that evening.

During the evening hours, several residents living nearby passed the Columbine grounds and heard the sounds of a lively party inside. They testified later that the noises within sounded very much like a drunken revel. Scherf's host, Columbine mine superintendent Frederick "Ted" L. Peart, chose this weekend to send his family away for a few days.

Merle Vincent and a number of his associates became apprehensive when they learned about the buildup of state military at the mine; Vincent went to visit Governor Adams at the Brown Palace Hotel in Denver. Vincent wanted to know why Scherf and his men were at the Columbine. The governor was unaware that Scherf's force was indeed in the northern field; he told Vincent that he had not given such orders for the men to move north. The governor seemed to realize the seriousness of the situation and told Vincent that he would contact Scherf and order him to depart the Columbine.

Apparently, Louis Scherf was always out when the governor phoned the mine. For some reason, he was unavailable, even after midnight. Now the governor, becoming apprehensive himself, sent Thomas Annear, chair of the Industrial Commission, the man who had called the strike illegal, and Lieutenant Governor Sterling B. Lacy by auto to the Columbine to investigate the matter. It was now 2 A.M. or so; the governor's men set off sometime later. They were motoring up the road to Serene just as dawn was breaking, about 5:30 A.M..

Young miners Lawrence Amicarella and Jimmie James were also on the road from Denver at that hour. They had spent a Sunday night "on the town" and were about to enter Serene from the south when they were stopped and checked by a Columbine guard nicknamed "Pork Chop," who habitually wore a large cowboy hat. Amicarella went home to his family's company house inside Serene and got ready for bed.[1]

As the sun came up, some one hundred autos brought the picketing men, women, and children within two hundred yards of the outside gate; others came on foot. Soon, from six hundred to one thousand pickets—estimates vary widely—formed their ragged lines outside the fence at the north of the Columbine.

Breaking into song, their flags held above the crowd, the strikers and their families started to move. Jessie Irwin and her friends had parked their Oakland auto and ran excitedly down the dirt road to join the others.

The sun was up. So were Louis Scherf and his men. Columbine guard "Cowboy" Gene Ewing had climbed the tipple and had a gun—was it a carbine or a machine gun?—at the ready. One of Scherf's men was high on the water tank with a visible machine gun trained on the inside gate.

I.W.W. organizer Adam Bell led the lines toward the first gate. Beside him was faithful marcher Elizabeth Beranek—Joe and the children were somewhere in the lines—and she had her large American flag firmly held high this morning. She and Bell led the group toward the first gate, which was closed but not locked. In they went. Scherf's men were grouped behind the second closed gate. Scherf himself was not in uniform and therefore not easily identified by the strikers. Later, he reported that he ordered the picketers to stop moving; he added, "If you come in here, you will be carried out."[2] Obscenities were exchanged. Another guard is alleged to have shouted, "Watch out, boys, my fingers are getting itchy."[3]

Strikers mill about in front of Erie doctor's office waiting for news of the wounded. *Denver Post*, November 11, 1927.—*Photo, Western Historical Collections, University of Colorado at Boulder*

When Bell attempted to climb the gate, he was pulled over inside the barricade by the guards and clubbed to the ground, where he lay unconscious and bleeding with head wounds. Mrs. Beranek tried to cover Bell with her flag; it was wrenched from her hands, the staff broken and the flag ripped up. She was beaten as well.

There may have been some hand-to-hand fighting between striker and guard. Rocks may have been thrown at the state men. At this point, Scherf ordered his men to throw tear gas grenades into the crowd; the strikers responded by picking up the bombs and hoisting them back to the guards.

As the picketers continued to surge forward—Scherf said later he fired a warning shot in the air—the guards started to fire handguns, rifles, and the machine gun. "Let 'er go."

With the first shots, Lawrence Amicarella, who was getting ready for bed, dropped his shoes and went for the floor. Mary Borstnick and her family, who lived next door to the Amicarellas, all hit the floor as well. Jessie Irwin and her friends turned away from the road to Serene and ran as fast as they could toward Lafayette and their car.

Thirty-four-year-old striker John Eastenas, a Lafayette father of six, was the first to fall; he died immediately. The miners scattered. Mary Borstnick remembered hearing Elizabeth Beranek run past her home, shouting, "Those sons-of-bitches. They killed him"[4]

In addition to Eastenas, Lafayette bachelor Nick Spanudaklis, thirty-four, a striker from the Morrison mine, was fatally wounded and died a few hours later in an Erie doctor's office. Jerry "Jay" Davis, a twenty-four-year-old miner from

The stunned family of John Eastenas poses for a Denver cameraman one day after he was shot down at the Columbine, 1927.—*Photo Thomas Costigan scrapbooks, Western Historical Collections, University of Colorado at Boulder*

Frederick, Rene Jacques, twenty-six years old, from Louisville, and George Kovitch, twenty-five, from Erie, died from their wounds soon after. Russian-born Mike Vidovitch, thirty-five years old, a veteran of World War I, who was working at the Clayton mine, died a few days later in a Longmont hospital.

Sixty pickets, both men and women, were seriously wounded or otherwise injured; three guards were hurt. Scherf and his men later claimed that several sharpshooters were in the crowd of strikers, which caused the military to fire at the lines. Someone reflected that the miners must have been lousy shots if Scherf's statements were true. No one inside the Columbine compound was shot.

STICK WITH THE PILOT!

Cartoon from the *Rocky Mountain News*, November 21, 1927.—*Western Historical Collections, University of Colorado at Boulder*

Colorado National Guardsmen stop car on road near Serene.—*Photo,* Rocky Mountain News, *Western Historical Collections, University of Colorado at Boulder*

By that evening, Governor Adams had apparently swung into action. He backed up Louis Scherf and his men without question. He listened to reports from Industrial Commissioner Annear and Lieutenant Governor Lacy. "I do hereby find that tumult, riot and mob violence are threatened in the counties of Boulder and Weld . . . and I hereby find and declare that a state of insurrection exists in said counties."[5]

Adams finally ordered out the Colorado National Guard. A force of 325 men, supported by 45 cavalry and two whippet tanks, was quickly in place. Airplanes continued to patrol the area from above. Guardsmen were also posted to the Puritan, Grant, and Monarch mines. A 10:30 P.M. curfew was imposed; passes were required to get in and out of the mines. Among those taken to the Longmont jail were Adam Bell, Tom Harris, George Kubick, and James Allender.

At 26 Broadway in New York, the offices of John D. Rockefeller, Jr., were also picketed with such quickly painted slogans as "Down With the Colorado Cossacks."[6] On November 23, the *New York Times* reported to its readers: "With the strikers remaining in their homes, the little coal mining villages of Erie, Lafayette, and Frederick took on a deserted look, as hardly a person appeared on the streets, where for the last few weeks the miners have gathered daily to discuss the strike situation."

No matter that the Wobblies organized the strike. All Colorado labor groups, including the State Federation of Labor, no longer held back but expressed their outrage at the Columbine killings. Numerous church groups took a stand against the Columbine guards and state military and their actions. Journalist Frank

Palmer was later to write in the *Nation*, "We find that the present devices of our State to prevent discordant strikes have broken down."[7]

While Lafayette residents mourned the death of John Eastenas and the others, a service for those who died at the Columbine was also held at Ludlow, where, fourteen years earlier, others had died as a result of similar industrial warfare.

After a Weld County coroner's jury headed by Dr. John A. Weaver heard seven hours of testimony on the deaths of John Eastenas and Nick Spanudaklis, its verdict, delivered on November 23, after thirty minutes of deliberation was:

> That the said deceased parties . . . came to their deaths in Weld County, Colorado, on the 21st day of November, 1927, as a result of gunshot wounds received in and upon the property near or at the Columbine Mine while they were *en masse* assembled and invading the mine property, after having been properly and legally warned not to enter the same and of the perils attending such invasion by the proper and lawful officers of the State of Colorado; that said gunshots were fired by persons unknown and who were members of the state law enforcement department, and said deaths are not felonious.[8]

That same day, a letter of petition was presented to Governor Adams, signed by six thousand Colorado coal town residents, asking that he set up a meeting with strikers and operators represented. The goal? To end the strike. Adams did attempt to schedule such a meeting but the operators said they would not come. Coal field businessmen also tried their collective hand at mediation and visited with miners' groups to determine if a solution were at hand. This time the miners joined the operators in refusing to submit to mediation, saying it was a ruse to get them back to work without proper redress of their grievances. Nobody, of course, wanted to meet with the Wobblies.

The I.W.W. did offer food and supplies to the striking families in the northern field, which were grudgingly accepted by some and refused outright by others.

Funeral march for John Eastenas, killed at the Columbine. *Rocky Mountain News*, November 24, 1927.—*Photo, Western Historical Collections, University of Colorado at Boulder*

Finally, a number of northern field operators had a change of heart and agreed it was time to sit down at the table of the Industrial Commission. Now it was the commission's turn. The members voted not to mediate the strike. After a few additional requests, however, the commission changed its mind and voted on December 10 to hear testimony beginning December 19.

The commission's report, which came out on March 20, 1928, addressed the question of the surprising strength of the I.W.W. during this period and concluded that systematic wage cuts, continued abuse in weighing coal by company weigh bosses, lack of compensation for dead work, foul air and dangerous conditions underground might have had something to do with miner discontent. Maybe, the commission's report went on to suggest, just maybe, both worker and operator should consider collective bargaining to achieve positive change.[9]

Meanwhile, the behavior of the national guard was under scrutiny. At a Longmont evening meeting on December 15, after Frank Palmer and six others had completed a session of fiery speechmaking, the guardsmen were ordered to arrest them. The military entered the meeting hall, placed the seven speakers under arrest, and marched them out, with bayonets pointed at a frightened audience. Longmont businessman Thomas F. Mahoney was at the meeting and was outraged that any state military man would point a bayonet at him. He said to a reporter:

> Longmont is a quiet and orderly farming town and the meeting Thursday night was orderly and the temper of the crowd good-natured. Yet we were held in the hall with locked doors and then marched out between two lines of militiamen with fixed bayonets.[10]

Mahoney added:

> I do not believe Governor Adams is in sympathy with this sort of thing, but I think he is being misled by his subordinates in the militia.[11]

Mahoney also stated that it was no wonder to him that miners responded with violence if the state's national guard behaved in such radical fashion. Mahoney was not the only critic of the guard's actions in Longmont. The governor's office was besieged with letters, telegrams, and phone calls on the subject. The Denver Typographical Union, Local 49, had this to say:

> It appears that Mr. Palmer had made a speech to a meeting of Longmont citizens in which he freely criticized the chairman of the Colorado Industrial Commission, Thomas Annear, and officers of the guard . . . if it is a crime to have a poor opinion of the amiable incompetent who heads the Colorado Industrial Commission, please count the members of our union all as criminals.[12]

A Denver paper allowed that Governor Adams "Was Between Two Fires Directed by His Personal Friends":[13]

> There had been the faction which demanded that the strike be crushed and discredited at all hazard, the "line 'em up against the wall" group. . . .The strike breaking group is making its last stand in the state capitol. . . .At several times the governor's orders in regards to management of the strike have been ignored by officials who belong to the strike breaking faction and he has been left to deal with the unforeseen consequences of acts which he had definitely forbidden.[14]

Did someone in Billy Adams's cabinet order Louis Scherf to the Columbine without the governor's knowledge and consent? Whatever the case, the following Saturday night, Adams announced to the press: "I am through with politics. I will not be a candidate for governor in 1928. Neither will I again be a candidate for my old seat in the senate or for any other office"[15] Billy Adams had had enough.

As she tried to gain full control of the Rocky Mountain Fuel Company, Josephine Roche was distraught over the events at the mine at Serene. She publicly expressed her sorrow over the deaths of the striking miners. She did not hide her anger with Governor Adams and his methods, or lack of methods, to quell labor disturbances in the northern field. Privately, she wondered aloud to her aides about the rumors that officials of the C.F. and I. to the south had somehow "encouraged" the Wobblie movement in Colorado. She also wondered who ordered Louis Scherf and his men to move so quickly from Walsenburg up north.

Miss Roche engaged the services of private detective J.B. McQuarrie and others to interview a number of parties involved in an effort to get at the truth of the situation. Did the governor send Scherf to Serene? If not, who did? Were any Rockefeller men involved in the events leading up to the Columbine massacre? Did they pay the Wobblies to come into Colorado? Did they promise something to Louis Scherf that caused him to move his men without the governor's knowledge?

Detective McQuarrie was busy traveling about the West for a number of months following events at the Columbine. He found Adam Bell in Montana. Bell discounted any connection between the I.W.W. and officials of the C.F. and I.[16] Louis Scherf told McQuarrie that he had been "used" by the Rockefeller firm.[17]

In January 1928, spinal meningitis spread through the coal towns in the northern field. Casting aside logic, a number of the miners bitterly accused the coal operators of poisoning the wells throughout the field to cause spread of the disease.

Wobblie A.S. Embree spoke to a group of miners on February 20, congratulating them on their "partial victory."[18] But, he said, it was time to go back to work. Some 1,700 northern coal miners had already returned to their jobs. A number of the men, however, were still feisty. Embree told them, "Do not mistakenly make . . . sacrifices of no avail by refusing to return to work now. . . . Vote for your organization. Vote to return to work at once. Take your splendid solidarity with you to your jobs."[19]

The following day, the last of the prisoners were released from the Longmont jail. Several days later, national guardsmen left the Columbine mine and the other properties at which they had been installed. At the Columbine, Miss Roche prevailed upon her board and ordered that Ted Peart be "released" from his superintendency at Serene; Peart left immediately. Governor Adams recalled and deactivated Louis Scherf and the men of the state law enforcement department. Before Adams completed his term, he ordered that Pat Hamrock be removed as a prison chief at Canon City.

Suddenly, as quickly as they had entered the state, the "foot-loose rebels" departed. By the end of March, only a few Wobblies remained in Colorado to celebrate the I.W.W.'s "last great fling."[20] Somehow, the I.W.W. leadership knew their strengths—rallying laboring men to a cause—but they also sensed that maintaining regular union activity was not their forte.

A.S. Embree, upon reflecting about labor organization in general said, "The end in view is well worth striking for, but in the struggle itself lies the happiness of the fighter."[21] While the Wobblies were there, it was quite a strike.

Historian McClurg reported two incidents concerning the Wobblies that local marshals did not forget quickly:

> In Erie, the marshall is said to have been beset by dissatisfied and angry deputies. Out-of-state Wobbly organizers imprisoned for strike activities had convinced their guards that jailers in Utah and Wyoming enjoyed higher pay, shorter hours, and better conditions. The prisoners had advised the formation of a Deputies Union.
>
> In Lafayette, a group of fifteen imprisoned strikers had organized a jail committee, set up classes, and devoted their time above all else to making the marshall's life miserable by incessant singing . . . refusal to follow jail routine, etc. When he finally capitulated and told them to go, the strikers solemnly held a committee meeting to consider his proposal. The committee's decision was that, since the marshall would only replace them with more picketing miners and since they were already acclimated to life in jail, they should remain. It was a full week before the marshall got his jail back. It is said that he imprisoned no more pickets.[22]

During the strike months, a student of the period, Hazel Alice Glenny, lived in Lafayette for a time during the preparation of her master's thesis on the subject. She talked to many of the striking miners:

> Many who were out on strike expressed opposition to the leadership of the Industrial Workers of the World and refused to accept the financial aid it rendered. Despite this fact, their feeling of loyalty to their fellow workmen and their opposition to the operators made them select the strike rather than remain at work. The feel of class consciousness was highly developed among the miners, particularly among those of British and American descent, and although they may have taken no active part in the strike except to remain away from work, a return to the mines was looked upon as traitorous.[23]

Glenny remembered visiting with one miner who had chosen to go back to work in January, in advance of the strike settlement; he told her that

> his father would rather he died than enter the mines before the strike was officially terminated. His mother, in actual need, refused to accept his earnings. His friends turned against him, and he became practically an outcast.[24]

Another historian of the I.W.W. had this to say:

> Some may say the strike was too long, others that it was too short, but none can deny that it was pulled at the right time and that the tactics pursued were proper as well as new to the industry. . . .Considering the number of people involved in it, it was the cheapest strike ever held in this country from a financial viewpoint. Yet, more was accomplished than in any other five strikes in the State.[25]

A "modern" tipple at the Gorham mine. "Lump" or quality coal is at the right; "egg" sized coal is lower grade; "slack" coal was poorest grade, sold to industrial firms requiring steam.—*Photo, Boulder Historical Society, Carnegie Branch Library for Local History*

CHAPTER FOURTEEN

In March 1928, Josephine Roche was finally able to gain control of the forty-year-old Rocky Mountain Fuel Company by the purchase of Denver real estate man Horace W. Bennett's holdings. Now that she had 51 percent of the firm's stock, she called for a new election of the board of directors, elevated Merle Vincent to president, and hired former labor organizer John Lawson to serve as vice-president. Miss Roche became a vice-president as well. Former union lawyer Thomas P. Costigan agreed to serve as general counsel for the coal company. (Costigan would become Colorado's United States senator in two years' time.)

After inspecting the company's general offices, Miss Roche cleaned out the guns and other forms of armament stored there. She reviewed and put aside those company records that showed immense expenditures for guards and detectives. She saw that some officials who had not been purged immediately after the Columbine affair "walked the plank," as some of the miners put it.

The new officers felt it was time for the unionization of Rocky Mountain Fuel mines, but they weren't overly pleased to work with officials of the Industrial Workers of the World. Costigan met with the few Wobblies who were left for a time, but he finally withdrew from negotiations with them.

When Rocky Mountain Fuel contacted officers of the United Mine Workers to propose a two-year contract, it was a shot in the arm for the labor group. A revitalized U.M.W. group met with Miss Roche and her officers in May 1928. By June, three-fourths of the northern field was organized by the U.M.W. The contract was signed on August 16 and was to take effect on September 1. It was a most unusual contract and probably gave other Colorado operators apoplexy when they read its preamble:

> We, the signers of this document, seeking a new era in the industrial relations of Colorado, unite in welcoming this opportunity to record the spirit and principles of this agreement.
>
> Our purposes are:
>
> To promote and establish industrial justice;
> To substitute reason for violence, confidence for misunderstanding, integrity and good faith for dishonest practices, and a union of effort for the chaos of the present economic warfare;

> To avoid needless and wasteful strikes and lockouts through the investigation and correction of their underlying causes;
>
> To establish genuine collective bargaining between mine workers and operators through free and independent organization;
>
> To stabilize employment, production, and markets through cooperative endeavor and the aid of science;
>
> To assure mine workers and operators continuing mutual benefits and consumers a dependable supply of coal at reasonable and uniform prices;
>
> To defend our joint undertaking against every conspiracy or vicious practice which seeks to destroy it; and in all other respects to enlist public confidence and support by safeguarding the public interest.[1]

Rocky Mountain Fuel employees would work an eight-hour shift six days per week with eight paid holidays per year. The minimum daily wage of $7—Rocky Mountain Fuel miners had been receiving $6.77—was the closest Colorado miners had come to the 1924 Jacksonville scale. There would be two paydays per month, a checkweighman at each mine, and pit committees to handle grievances. The men who went back to work for Miss Roche were assured they would not be penalized for their temporary "zeal" in working with the I.W.W. Rocky Mountain Fuel employees were now the highest paid in the state.

It seemed that the promise of good times ahead might be fulfilled after all, despite the growing depression nationwide and the further introductions of natural gas—Sinclair Oil Company, Standard Oil of New Jersey, Public Service Company of Colorado—into the Denver market during the summer of 1928. Miss Roche and her associates were trying for reason and order in an industry that had been characterized as "the most disorganized in the United States."[2]

From a strike to a depression. The financial hard times of the 1920s melded with the financial hard times of the 1930s in the northern field. After Wall Street "failed," coal mining families did not find it difficult to adjust to a slimmer pocketbook; most had been on short rations for a number of years.

No sooner had the men gone back to work but the summer slack season emptied the mines of most of its workers again. The men came home to tend their vegetable gardens. For many, it was all the food they had. Before the Colorado National Guard left the area for good, some lucky young boys were "adopted" by the military men, who saw to it that the lads had one good meal per day.

Some of the men went out to the sugar beet fields each day "to work the beets," bringing home from two to four dollars per acre. The women worked outside the home, too. Both men and women worked night shifts to "top" the sugar beets, readying them for Great Western factories. Some snipped green beans for ten to fifteen cents per hour. At a factory in nearby Brighton, the picking, peeling, and chopping of tomatoes brought fifteen cents per bushel. Lafayette was the site of a pickle factory, located on east Emma Street; some found work there.

In 1933, Mother Nature contributed to hard times with a massive grasshopper plague, a cyclical event in eastern Colorado, which decimated that season's gardens and fields.

Each day, northern field children policed the railroad tracks to stuff their gunny sacks with chunks of coal that had tumbled from loaded freight cars; they lugged the sacks home for the family cookstove. When Superior children were out looking for pieces of coal, they often carried a potato in each pocket; if they chanced upon a still-smoldering slag pile, they roasted their potatoes in the coal fire for a special treat.

In Lafayette, Mayor Robert Johnson signed for those families who qualified for a government-issued bag of groceries. Elizabeth Beranek tried the federally sponsored food supplies for a time but discovered that the flour that was issued wouldn't rise enough to produce bread for her brood of seventeen children.

The men would go on hunting parties from time to time to bring in freshly killed rabbits. Jessie Irwin James still vividly remembers the row upon row of dead rabbits, stretching along downtown sidewalks, waiting to be picked up for the family stewpot. Now and then, a soup line was established on Simpson Street.

Prices at local company stores came down a little. A new pair of overalls might cost $1.49, a shirt cost fifty-nine cents, and a pair of work gloves cost ten cents.

As elsewhere in the country, not all was grim. Lafayette's Jewel Theatre was well patronized; one could see a silent movie for six cents. Jessie Irwin played the piano as background to the silents until the sound came with the films. Miners and their families still went to union-sponsored dances and moved about the floor to the music of a violin and piano. Band concerts still drew the crowds. Baseball commanded an almost total civic attention, with local merchants closing down for an important game; some games were played on Sunday afternoons.

Saturday nights found a crowd at Lafayette's company store; manager W.F. Autrey would turn up the radio so that no one would miss a second of the fights. Each coal town's hose team continued to practice its skills even though some of the camps could now afford modest motorized equipment. There hadn't been a state

Out for a summer ride.—*Photo, Louisville Public Library*

Carlo Damiana and young Bob Overland, 1924.—*Photo, Louisville Public Library*

Sledding in Community Ditch near Superior.—*Photo, Winston Morgan*

championship won locally since Lafayette took the honors in 1924.

After seeing reports in the newsreels about the Olympic Games, a group of young men, Boughton "Boots" Noble, John "Greekie" Lewis, and Ponzo Mastriana devised their own athletic events with materials at hand. They "threw the discus" with a used stove lid. Their "javelins" were dried horseweed stalks, which hardened into perfect throwing devices.

On a hot summer afternoon, the boys might skinny-dip in Coal Creek or catch frogs along its banks to roast over a small fire. At night, they climbed the fence surrounding Public Service Company property to swim in Weisenhorn Lake, once a resort area. The burros from the Eldorado Springs resort, released to graze during the warmer months, were chased by the young people, hoping for a free ride.

Erie children played in "dugouts" or caves near town, which, they were told, had existed since the days of the Indians, before white settlement. They had to be careful, however, because some of the caves were occupied by displaced families. Everyone watched for the bands of gypsies that would wander mysteriously through the area.

In winter, Superior children skated and sledded down the icy Community Ditch.

With the Denver and Interurban into receivership, residents without autos weren't able to take as many expeditions to Denver, Longmont, or into the mountains as they had once done. Few families could afford the Oakland, Regal, Star, or Ford automobile. By 1932, Frederick, Firestone, and Dacono residents could no longer take the train to Denver. Year by year, passenger routes on other lines declined, then were dropped altogether.

In 1924, the bank in Lafayette was robbed. The thieves blew apart the bank's safe to get at the five- and ten-dollar gold pieces stored within. In their flight, they evidently dropped several sacks of coins, which were scattered about the area. For months, even years, young children would bring in their recently found "golden penny" in hopes it would buy some candy.

With the depression came the Works Progress Administration. The W.P.A. projects brought improvements and a few jobs to the coal towns. Erie now had dikes constructed along the banks of Coal Creek to protect the town from its periodic flooding.

Lafayette got a new waterworks as well as sewer and street improvements. The town also contracted with the federal government to build a community swimming pool, a project that would divide the town and cause hard feelings for years to come. During the spring of 1933, teams of horses pulled the trenching equipment back and forth to dig out a long, deep rectangle for the pool. After it was lined, the new pool opened for business under the management of the Lafayette Fire Department. Near the entrance of the pool was a posted sign that read: "Firemen's Pool. We reserve the right to reject any or all persons without cause. White trade only."

Since many local citizens did not regard their Spanish-speaking neighbors as members of the Caucasian race, everyone knew what group was meant to be excluded. No matter that their taxes, too, had gone to build the pool. Mexicans were diseased, some said, unfit to swim with the "white" children. A group headed by Rose Luceros, through their Denver-based attorney, filed suit in district court over the pool's restrictions, saying that the pool, which was built on City of Lafayette ground, must be considered a "public accommodation and amusement."[3] The suit further stated that such continued restriction might well lead to "violence, disorder and peace of the town."[4]

By the following summer, District Court Judge Claude C. Coffin "in a double-edged 'empty victory' . . . denied the injunction of the Spanish-American residents of Lafayette to force the members of the fire department of that town to let them swim in the municipal pool."[5] The judge ruled that the document wasn't properly drawn. He also said, however, that Mexican children could not be barred from swimming in a tax-supported pool.

City officials responded to the court's action with the radical decision to bulldoze tons of dirt into the pool to cover the recreational site. No more swimming pool. One Lafayette resident, in reminiscing about such municipal activities, laughed, and said, "You know, to this day, I think how selfish can people get. You know, they had it all ready, it was all ready, but they had a quarrel over it. They didn't want the Mexicans in it or something. And how foolish. I don't know what difference it would make."[6]

Miss Roche and the miners at Rocky Mountain Fuel continued to flourish for a time despite the depression, the installation of more natural gas lines, and the relentless opposition to her policies by other coal operators in both fields. Now

Lafayette youngsters pose for school photo.—*Photo, Lafayette Public Library*

Main Street, Louisville, looking north, 1923. At right is LaSalle Pool Hall.—*Photo, Louisville Public Library*

Rocky Mountain Fuel holdings were second only to Rockefeller's Colorado Fuel and Iron Company.

In August 1931, C.F. and I. received permission from the Industrial Commission to cut wages at its mines to $5.22 per day. At the same time, the giant coal operator lowered its coal prices as well. This was a blow to Josephine Roche and her business expectations. But something happened that may be regarded as unusual in any American business. Rocky Mountain Fuel miners met to discuss the possibility of lending Miss Roche some eighty thousand dollars by accepting only half of their regular pay for a 2½-month period. The agreement to help out Miss Roche was not compulsory; not every miner in the field was able or so inclined to join in the loan effort.

Miss Roche's competitors immediately accused her of sneaking in a pay cut by ignoble means despite the fact that the company repaid all loans from its miners a short time later. The miners established a Coal Sales Committee, which encouraged prospective customers to "Buy With Josephine." As one union official said, "We don't want another strike like that in 1927 and you don't want to pay the cost for another one either."[7]

As more and more natural gas was piped into Colorado, less and less coal was sold for household or industrial use. In May of 1932, officers of the C.F. and I. came again before the Industrial Commission to seek another pay cut—15 percent—for its employees, 695 of whom had petitioned the commission for redress. This time the Industrial Commission stood up to C.F. and I. officials and denied the proposed wage cut, stating that "the men are not receiving a living wage under the present scale."[8] C.F. and I. defied the commission and lowered its wages anyway. Historian Robert Athearn, in later years, described the company's actions as "industrial slavery."[9]

By this time, Rocky Mountain Fuel employees were earning less as well—from $7 down to $5.25 per day. As sales continued to drop, Miss Roche devised an employment system whereby the work was divided so that a greater number of men could bring home a little bit of money to their families. She also directed the management of her company stores to offer credit to those employees on hard times. She saw that house paint and other property improvement supplies were available at little or no cost to miner homeowners. No wonder miner Lawrence Amicarella called Miss Roche a "good old soul."[10]

With the establishment of the National Industrial Recovery Act in 1933 under the auspices of the Franklin D. Roosevelt administration, all coal mines in the United States were automatically unionized. Thus, after sixty-two years of effort in the northern field, unions for coal workers became a federal law.

After this time, Miss Roche was in and out of the management of the Rocky Mountain Fuel Company. When friends, including U.S. Senator Thomas Costigan, urged her to enter the political arena, Miss Roche declared herself a candidate for the Colorado governorship in 1934; she lost a particularly bitter Democratic primary election to the incumbent, Edwin D. Johnson.

She barely had time to shed a tear at her failure to win, however. The president of the United States sent word that he wanted her to come to Washington, D.C., to take over the post of assistant secretary of the treasury. She accepted the position and left Denver to supervise the Public Health Service as well. Later, Miss Roche directed the National Youth Administration, one of Franklin Roosevelt's many alphabetical agencies. Josephine Roche was regarded as the second most important woman in Roosevelt's administration; the first was, of course, her old college friend, Frances Perkins, secretary of labor.

Miss Roche was back at the helm of the coal company in 1937; however, production had decreased 23 percent. Again she was helped by the miners. This time, her friend John L. Lewis, still president of the United Mine Workers, authorized loans amounting to $450,000 issued through Lewmurken,* a union "dummy" corporation. Miss Roche also borrowed money from other friends, sought help from the federal government, and invested much of her personal fortune into the ailing company.

All to no avail. The Colorado coal industry was not merely ailing; it was dying. In 1928, 9,934,000 tons of coal were mined and sold in the state. In 1939, only 5,722,899 tons of coal were shipped to market. That June, John L. Lewis sent out to Colorado an Ohio firm, Coal Mine Management, to take over Rocky Mountain

---

* Lewmurken was the combination of the names of John L. Lewis, U.M.W. vice-president Philip Murray, and secretary-treasurer Thomas Kennedy.

Fuel from Miss Roche and John Lawson. Thus, "the once highly touted social experiment conducted in the Colorado coal fields had run into the rocks."[11]

In the early morning hours of January 20, 1936, the day shift—about 150 men—at the Monarch No. 2 mine, located two miles north of Broomfield and two miles south of Louisville, was gathering around the elevator shaft waiting for the cage to come up and get them.

Down below, ten miners were completing the graveyard shift, spread out in twos and threes along the miles of underground tunnels. The noise of their work echoed down the passageways. Young Bill Jenkins, a novice miner, was working with older, more experienced Nick Delpizzo away from the center of the mine entries.

Suddenly, in the quiet chambers, Delpizzo looked up. He could hear what Jenkins could not. Something was wrong. "Mine explode!" Jenkins turned to the older man. "What to do?" Delpizzo shouted, "We gonna run! If not, we'll die on road . . . between life and death . . . very short . . . can choke up . . . people will find us dead. . . .We gonna run like rabbits!"[12] And run they did. They tore down a four-thousand-foot tunnel, turned to reach a ladder, and climbed the rungs for three hundred feet up to safety.

On top, the waiting shift felt the blast beneath their feet and estimated it was a big one coming from the southwestern section of the mine. A rescue operation was

Waiting for news at the Monarch No. 2, January 20, 1936.—*Photo, Boulder* Daily Camera

immediately mounted, but the drifting methane gas and coal dust made the job difficult and dangerous. Eventually, the bodies of Tom Stevens, Louisville; Ray Bailey, Broomfield; Oscar Baird, Richard's Camp; Tony DeSantis, Louisville; Steve Davis, Louisville; Kester Novingger, Broomfield; and Leland Ward, who lived at the Monarch mining camp, were found.

Some of the men were discovered seated on the floor of the mine, their equipment at their side, their faces turned toward their lamps, as if to watch the flame burn out with the creeping presence of carbon dioxide, the black damp.

The rescue crews searched for the tenth miner, Joe C. Jaramillo, whose family lived in company housing at the mine. After eleven days, officials of the National Fuel Company, owners of the Monarch No. 2, offered Mrs. Jaramillo one thousand dollars if she would give permission for the rescue crews to stop their work. A different company attitude from that of 1890 or 1910.

Mrs. Jaramillo was loath to give that permission because she felt Joe had gone down below with a money belt strapped around his waist; she knew he could no longer be alive. (Later, the nine hundred dollars was found safe in a nearby bank account.) Joe Jaramillo was never found. Some of the mules survived and were brought up to safety.

Later, when the accident was investigated, it was determined that the coal dust,

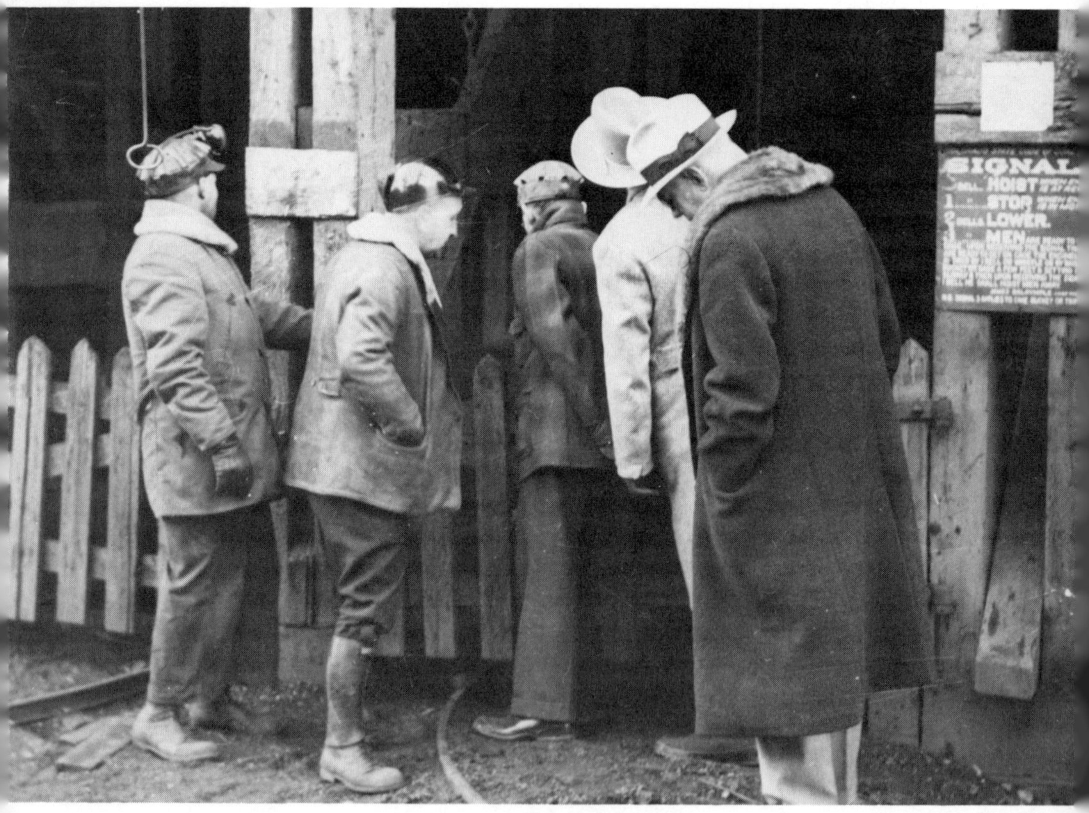

The cage is coming up.—*Photo, Boulder* Daily Camera

The news is bad.—*Photo, Boulder* Daily Camera

which lay five and six inches thick on the mine floor, had combined with a pocket of methane gas and was touched off by a spark, perhaps from a trolley. The operators were found negligent this time, a rare decision in the Colorado coal fields.

The company erected a memorial to the life of Joe Jaramillo. A five-foot slab of red granite was placed three hundred feet above where they believed Jaramillo lay entombed. The stone was inscribed, “Joe C. Jaramillo, 1887-1936, a faithful employee who died in the performance of his duty.”

Drivers on the freeway to Denver seldom see the Jaramillo stone.—*Photo, Boulder* Daily Camera

Superior is not much changed in 1978.—*Gary Rehn photo, Boulder* Daily Camera

CHAPTER FIFTEEN

In 1944, the Rocky Mountain Fuel Company declared bankruptcy. The once-powerful coal company had only one working mine and was four million dollars in debt. The United Mine Workers had lost at least five hundred thousand dollars because the union had helped Miss Roche. But "union leaders still remembered," that she had "opened the way to union organization of the Colorado coal mines."[1] One year later, the one-time heavy producer at Superior, the Industrial mine, closed its shaft. A good number of northern field mines—the Acme, the Sunnyside, the Vulcan, the Gorham, the Pluto No. 1—had already stopped operations at some point during the Great Depression.

Josephine Roche returned to Washington, D.C., in 1947 to become John L. Lewis's assistant; one year later, she became the director of the United Mine Workers pension fund, where she remained until shortly before her death in 1971.

With each succeeding year, more and more northern field mines closed. There was too much coal and too many miners for an era in which natural gas and electricity were the preferred power sources. In 1938, the federal government made a stab at a retraining program for out-of-work miners, but, in most cases, the unemployed "retrained" themselves.

Some miners made a living bootlegging whiskey until the end of Prohibition in 1933. Former miner Bauldie Moschetti went into the legal liquor business and flourished.

Michael Colacci, who had come from Italy in 1905 and worked in the Marshall mines, first tried the dairy business. In 1918, Colacci and his brother Jim opened the Blue Parrot, an Italian restaurant in Louisville. Through the years, other Colacci family members opened three more Italian restaurants in town. It became fashionable to drive to Louisville from many spots in northern Colorado to sample the Colacci family's cooking. (In 1988, the Blue Parrot burned, but Colacci family members were soon back in business.)

Many of the younger men opted for construction work or joined road crews, either for private companies or municipal governments. Some families continued to work long hours in the area's produce factories—sugar beet refineries, pickle companies, bean processors—until they too closed down in the 1970s.

In 1951, the Atomic Energy Commission decided that Rocky Flats, an area to the south of the coal field, was an ideal site for a highly secret facility using fissionable materials. First, the commission leased the new plant to Dow

Chemical, and triggers for atomic bombs were manufactured there. In 1975, Rockwell International took over the plant.

Scores of laid-off coal miners and their sons applied for work at the atomic facility despite the fact that many residents in the area were nervous about its presence along the populous Front Range. Over the years, groups have picketed the Rocky Flats plant to protest its location, its product, and its possible contamination of land and water nearby.

Many of the miners of the Northern Coal Field did not live out their days in comfort and good health. They were weak, coughed a lot, needed bottled oxygen, and had to sleep sitting up. They called it black lung. Some of the doctors called it pneumoconiosis, a disabling cardiopulmonary disease caused by the lodging of coal particulates in the lungs. Others cynically stated that there was no such thing as black lung, that those miners who smoked too much in their younger days were now paying for the bad habit.

When the federal government set up the Black Lung Benefits Program in 1969, the Social Security Administration was charged with the responsibility of handling such cases. Thousands of sick miners looked forward to receiving medical attention and financial aid for the unexpected medical bills. They made application for the benefits; only a relative few heard from the federal authorities.

The U.M.W., concerned about public opinion and its views on black lung, issued press releases showing that some four thousand coal miners died each year from the disease. Those doctors who pooh-poohed the existence of such a disease were usually on the payrolls of coal operators, said the union.

In 1973, the program was transferred to the Department of Labor in order that the coal operators might be made liable for paying black lung compensation. It was not surprising that the operators responded by fighting almost every claim for black lung benefits made by former employees. Exhaustive medical examinations became the rule for each retired miner who claimed disability due to black lung disease. The miner submitted to a breathing test, a blood-gas test, and a chest X-ray, as well as an examination by a coal company doctor.

The interpretations of these X-rays were often confusing and ambiguous. "It's just amazing how doctors can look at the same X-ray and arrive at different conclusions,"[2] said Edward Scheuneman, a lawyer for some northern field miners.

Former State Senator Harvey Phelps, a retired Army physician with a specialty in respiratory diseases, said that the basic problem with the Black Lung Benefits Program is that "it is an adversary battle instead of simply a pension program."[3] Phelps had been unhappy with the lack of response by federal agencies to coal miners' claims and with the operators' contesting each claim. "We became a rich, industrialized nation on the back of the coal miner. Untold thousands were killed, crippled or tossed out of work with disabling lung diseases like black lung and forgotten. . . .Now that's one helluva way to do things, isn't it?"[4] By the mid-eighties, 454,831 coal miners in the United States had filed for benefits. Some were successful; others were not. It seemed to the miners a willy-nilly lottery-type operation.

The little coal towns were changing. New people were moving in with no ties to the mines. They wanted a quiet and inexpensive place to live and raise their families. In 1950, Lafayette supported a population of 2,090; in the late eighties,

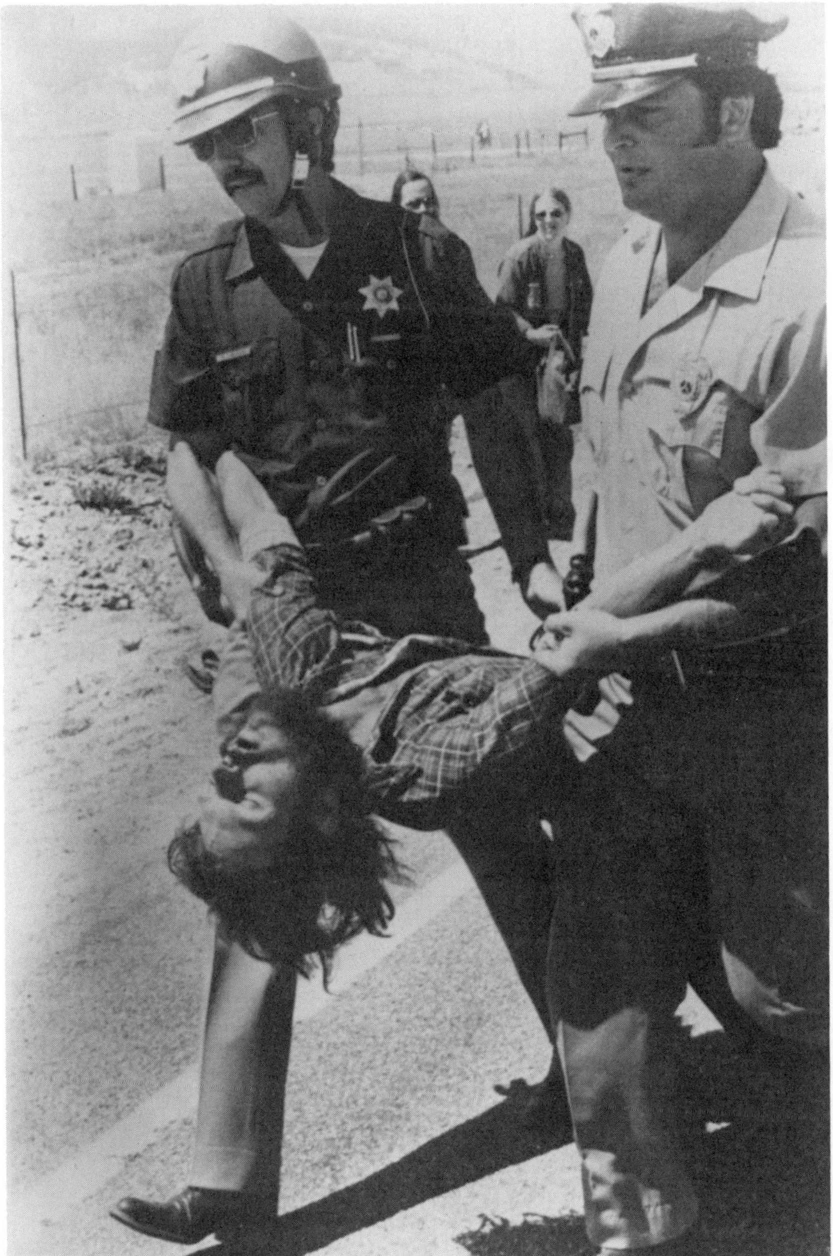

Police take away a demonstrator from the Rocky Flats property.—*Photo,* Town and Country Review

more than 13,000 people lived there. With home rule and the dial telephone came the developer. As was true in the other coal towns, Lafayette no longer had a solid financial base to its economy, nor could it depend upon a quality water supply. Often when the city council met to discuss possible expansion of the Lafayette city limits to accommodate a developer or a proposed improvement of its water services, reason was replaced by uproar.

A number of retired miners served on council and municipal committees and somehow got in the thick of it—an enjoyable pastime. Retired miner and councilmember Joe Mathias became so energized one December evening in 1970 that he was involved in a fistfight in council chambers. Fellow councilmember A.W. De Novellis felt compelled to write a long letter to the *Lafayette Leader*, complaining of Mathias's behavior and attitude:

> The flaring and pernicious remarks continually made by one of Lafayette's council members has a direct bearing upon the integrity and intelligence of every person in the city of Lafayette.
>
> The insidious remarks are directed to throw council in an uproar and to mislead the public into believing the present city council is nothing more than a flagitious group of pin-heads that should be ousted at the coming city election in September.[5]

After reading over his letter to the editor, writer De Novellis decided to add this note:

> It took me nearly one hour to look up in the dictionary all the words I wanted to use in this letter. This was not meant to confuse the public but merely to restrain me from losing my temper and using language such as—well—you know?[6]

Lafayette still has some of the old.—*Jerry Cleveland photo, Boulder* Daily Camera

A new Coal Creek Center at Lafayette, 1986.—*Vern Walker photo, Boulder* Daily Camera

Lafayette is going modern.
—*Vern Walker photo,*
Boulder Daily Camera

Ousting people from office seemed an important part of civic life in Lafayette. Over the years, police and fire chiefs were turned out, councilmembers recalled, and city managers fired. Sometimes for good reason. In 1969, one city manager resigned for reasons of health, or so he stated, but shortly thereafter, an inspection of the municipal books found a shortage of some forty-six thousand dollars.

For a number of years, traditional prejudices against people of other cultures in the northern field seemed to have coalesced around the Mexican-American residents. Incident after incident would bring such views out in the open for public attention. In the late 1980s, as Mexican-American residents took their places on municipal boards, committees, and city council, such prejudice paled in intensity compared to cultural conflicts in earlier years.

In Louisville, civic issues also concentrated upon water and residential annexations as well as shopping centers. In 1950, Louisville citizens numbered just 1,978; in the late 1980s, 9,327 people called it home. The regular removal of civil servants seemed popular in Louisville as well. In 1970, an over-ambitious police chief was put on probation by council because he raided city hall to find the mayor and friends engaged in a little gambling.

In the 1980s, some 250 Superior residents lived on the few dusty streets in town in modest homes, many crumbling and dilapidated, in the shadow of the Boulder-Denver turnpike, a commuters' freeway of cars roaring by. From its founding in 1896 to the mid-1980s, the little village had no sewer system or source of good drinking water. In 1983, the residents got their sewer, but the lines gave them trouble year after year. What few artesian wells were left in town were contaminated or nearly so; most families trucked in their water and bought bottled water for drinking.

Starting in the 1970s, occasional articles in the Boulder *Daily Camera* headlined a new and different Superior in the making. "Change is Near For Quiet, Country Town of Superior."[7] With a circuit rider city manager to advise them, members of the Superior Town Board pondered the possibilities for their little town. Shall we stay the way we are? Shall we welcome developers? Shall we annex enough acres to our town so that future homeowners will provide us with a source of water we can trust?

Angry words flew at the Superior Town Hall. Residents were violently split on questions that would affect Superior's destiny. Family members no longer spoke to one another. Recall elections were bitter and pitted one relative against another.

In 1987, Superior annexed 1,570 acres to its borders in hopes that developer Scott Carlson would provide them, at long last, with a decent waterworks. The three-square-mile land parcel, known as Rock Creek Ranch, would support eight thousand homes and a future population of twenty thousand. Not all residents were pleased. Said Eric Sampson, "Carlson wants to build a yuppie community. But he doesn't give a hoot about what's here today."[8]

Superior was not the only town in which recall elections provided a form of community "entertainment." In the 1980s, the little town of Erie experienced a recall election that divided its population of 1,250. Since the 1970s, Erie had annexed land to its borders, both in Weld and Boulder counties, with the same dream as Superior—funds for a modern waterworks that would produce decent drinking water. Sewers for Erie had been installed in 1956. The little town also

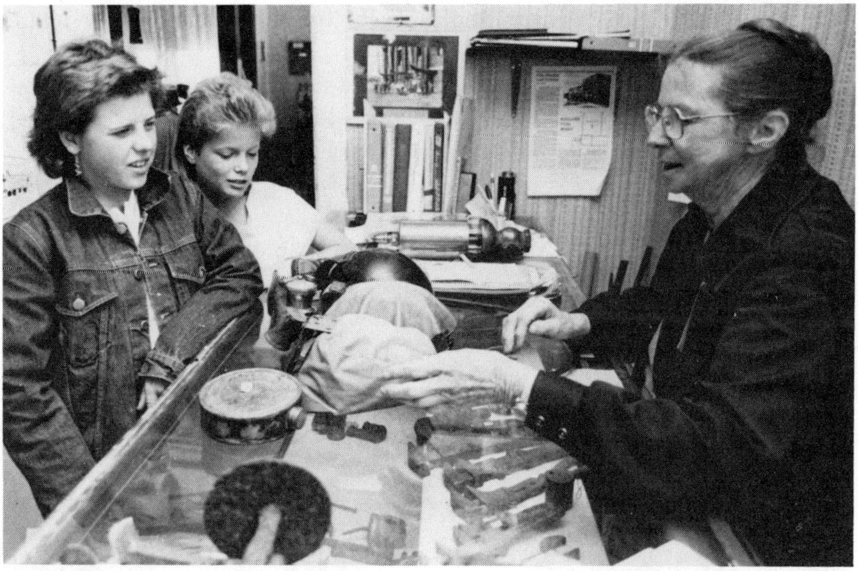

Ruthene Rodwick explains Lafayette Miners Museum artifacts to visitors, 1986.—*Charles Wendt photo, Boulder* Daily Camera

A tipple for the Black Diamond Saloon had to be imported from New Mexico.—*Vern Walker photo, Boulder* Daily Camera

used circuit rider city managers and planners to assess possibilities for future growth in the area.

Erie had an additional problem in that half of the community lay in the Coal Creek floodway and had experienced floods since 1876. Every few years, Coal Creek boiled over its banks, a threat to residents and property.

A reminder of Louisville's heritage.—*Vern Walker photo, Boulder* Daily Camera

Looking east at modern Erie.—*Photo, Boulder* Daily Camera

The airplane is a restaurant.—*Charles Wendt photo, Boulder* Daily Camera

In 1978, after a series of long civic fights, Erie annexed 834 acres so that the Erie Air Park might be developed. Contemplated was a small airport, an airplane-restaurant, surrounded by spacious and expensive homes. Someone predicted, "Erie will be like the Orange County, California, airport." Ten years later, Erie did have at least part of the air park, the hangars, and the restaurant, which opened for business now and again. A few "spacious" homes ringed the area.

In 1977, the National Oceanic and Atmospheric Administration built a one-thousand-foot tower nearby in order to study weather patterns moving off the Rockies; three times per week someone climbed this "Eiffel tower" to calibrate and repair the instruments.

In 1983, some of the residents near the airport mounted an unsuccessful campaign against the use of the ultra-light, a one-man, 254-pound motorized wing, at the airport. Their view was that the ultra-lights, which did not require a license to operate, were a threat to all who lived nearby.

Despite improvements in Erie, the trees planted along Briggs Street before the turn of the century were gone. The Eagle mine nearby, the last operating coal mine in the Northern Coal Field, closed in 1973.

To the north, Frederick also had water problems dating back to the days when the coal miners' wives used the water pumped from the mines for their laundry. During the 1950s, Frederick residents also battled with one another over the issue of $107,000 in water bonds. Impeachment of municipal officials was threatened. In 1954, the water bonds were defeated. Frederick did buy Firestone Lake, however, to use a reservoir. Three years later, a flash flood covered many Frederick streets with 2½ feet of water. In 1961, the flooding occurred again. Water they didn't need.

Growth issues surfaced in Frederick in the 1980s with its one thousand citizens fighting over a number of proposed annexations. Those in favor of a larger Frederick spoke of the need of additional revenue to build a modern waterworks in town. Elections featured hotly contested mayor and town board positions. In the background of political activity, citizens complained of continuing racist policies designed to thwart the ambitions of its Spanish-speaking residents. As in Lafayette, Mexican-American citizens chose to run for municipal offices and became part of the decision-making group in town.

In 1982, Frederick celebrated its seventy-fifth birthday. Yet, a few years later, there was talk of merging with its neighboring town, Firestone, and perhaps even Dacono as well. In January 1988, Frederick voters rejected the notion of a merger. In general, consolidation was favored by the younger citizens. The old miners wanted to keep the towns' identities separate. However, the three towns did cooperate in a common senior center, recreation district, sanitation district, and chamber of commerce.

In 1986, Firestone residents welcomed the reopening of Screwball Inn, earlier called Fishlein Hall, once a rough-and-ready institution that featured weekly dances, roller skating, silent movies, and regular fistfights between Firestone's farmers and miners. Thirteen hundred people made their homes in Firestone, but often worked elsewhere.

By the late 1980s, Dacono, the town whose name combined the names of Daisy, Cora, and Nora, had grown more than any other town in the Carbon

Valley. It now had 2,376 residents. Years before, in 1969, Dacono had annexed 480 acres to accommodate the development of the Glens of Dacono, a "suburb" of mobile homes. Near all three Carbon Valley towns, oil rigs had sprouted here and there to pump oil from the shale fractures below. Residents from all three Carbon Valley towns relied upon a nearby electronics firm for employment.

In 1973, the Boulder area was treated to a series of performances of *Columbine*, an opera in three acts. Mary Davis wrote the music and Joanna S. Sampson, long-time resident of the Marshall area, wrote the words to the production, which brought back memories of the days at Serene and the killings at the Columbine mine in 1927.

The following winter, the underground fire at Marshall flared above the ground for a time, sending flames from below four feet above the earth's surface.

Even though the Rocky Mountain Fuel Company was essentially a nonoperating organization, the concern made news in 1975 when its officials notified Marshall area residents that an un-named firm wanted to purchase mineral rights for a strip-mining project there. President Gerald R. Armstrong wrote to some of the residents that Rocky Mountain Fuel "is the owner of the mineral rights where your home is located."[9] "For a price of one per cent of the estimated current market value of your property, this company is willing to release such reservations by issuing a deed conveying the mineral rights except that of oil and gas."[10]

Denver lawyer Howard Potter appeared at a meeting of the Boulder City Council and told them he had a client who wanted the mineral rights to some 5,000 acres in the Marshall area. Opera-writer Joanna Sampson said she had gotten such a letter from Armstrong. So had almost everyone who lived in Superior. After a flurry of excitement over possible strip-mining in the area, the matter came to rest and was forgotten. And the Boulder City Council went ahead with its plans to purchase 1,200 acres of Marshall land for greenbelt acquisition.

Despite these periodic civic fights, the little towns of the Northern Coal Field seem to lie sleeping in the dry Colorado sunlight. Small weathered homes, a last vestige of company buildings, mark the older parts of the former coal camps. The tipples are gone now, with only a few pieces of grayed lumber to mark some of the entrances to the mine shafts and slopes. (When a Louisville restaurant decided to use the tipple as its main decor, the management had to send to New Mexico to import such a structure.)

Mary Borstnick, at ninety-four, widow of the mule foreman at the Gorham, still lives in a little white company house, dreaming of her happy days at Marshall—picnics, excursions, family holidays.

Steve Fotis, at ninety-eight, still remembers the pride of marching out of the Slope mine at Frederick in 1912 with fellow Greek miners, his friend Louis Tikas leading the way to membership in the U.M.W.A.

In Erie, Daisy Bracegirdle Swallow, at ninety-five, might well suddenly stand up in her sunny kitchen to sing:

> Hold the fort,
> For we are coming.
> Union men be strong.
> Side by side we battle onward,
> Victory will come.

In the winter, or on a wet, rainy spring morning, steam may still rise from unstable Marshall ground, evidence of the fire still smouldering below in empty coal chambers.

Retired coal miner John Craggs writes of his memories of the Northern Coal Field:

> In many coal camps today there is no evidence, no visible evidence, that coal was ever mined, but old miners like myself do not much remember the dirty miners' shacks, the company store, the hiss of steam from the hoist engines. What we remember and know are still there are the black chambers below the surface of the ground. . . .We remember that wide expanse of space at the bottom of the shaft or slope; it was supported with heavy timbers, and lighted with numerous electric light bulbs. . . .Here, mine cars of coal were pulled in by mules or dinky engines that they might be taken to the surface. . . .I have walked through the old workings. What a labyrinth of passage-ways!. . .It is incredible how well they were preserved.
>
> Today most of the buildings, the homes, the railroad tracks which were Marshall have disappeared, but underneath the rough terrain lies a monument to its hardy men who carved their own epitaph in the maze of under-ground tunnels and chambers. Here they did that which was necessary for them to live. . . .I often contemplate if in the distant future, someone will discover this real Marshall; this city underground that will still be the same as it is today even though thousands of years will have slipped by.[11]

For years the tipple at Marshall lay crumbling from rain and snow; it is gone now.—*Joanna Sampson photo, Boulder* Daily Camera

APPENDIX A
GLOSSARY OF MINING TERMS

BALDWIN-FELTS. A West Virginia-based detective organization that supplied Colorado coal operators with mine guards, detectives, and spies during the early 1900s.

BANJO. A scoop shovel used to load coal.

BANK. Early term for a seam of coal, usually an exposed outcropping.

BLACK DAMP. Deadly, odorless carbon dioxide gas.

BLACK LUNG. A disease of the lungs suffered by some long-time coal miners.

BINDLESTIFF. Pre-depression roamers who carried everything they owned wrapped in a small, blanket. Also *blankenstiff.*

BLOSSOM. Outcropping of coal.

BRATTICE CLOTH. A heavy canvas cloth that was hung near coal rooms underground and that was used to seal up a passageway to keep a fire from spreading. A *brattice board* served the same purpose.

CAGE. Lift or elevator in shaft.

CAGER. Worker who hooked loaded cars to be sent up to the tipple and dumped into railroad cars.

CARBIDE LAMP. The coal miners' light—it replaced the lard oil lamp—which was powered by a combination of calcium carbide and water.

CHECKWEIGHMAN. A worker, usually chosen by his fellow miners, who stood at the tipple to see that the operator weighed each man's tonnage correctly.

DEAD WORK. Those tasks that early miners were forced to do without compensation, i.e., timbering, laying track, cleaning up slag and rock.

DISTRICT 15. Area organization of the United Mine Workers of America which included Colorado, New Mexico, and Utah until 1922.

DRIFT. A mine tunnel level with the ground.

EGG COAL. A size of coal.

FIRE BOSS. A workman who preceded each shift into the mine to check for gas pockets and the like.

FOOT-LOOSE REBEL. Out-of-work Wobblies who roamed the country and who were readily available for picket duty.

FOXTOWN. Nickname for Marshall, probably derived from the nearby Fox mine.

GOB FIRE. An underground mine fire fed by waste products.

HEMATITE. A reddish iron ore found in the Marshall area.

JACKSONVILLE AGREEMENT. An agreement signed in February 1924 between operators in the Central Competitive Field and officers of the U.M.W.A. that established, among other matters, a daily minimum wage of $7.75.

JOHN MITCHELL DAY. April 1, 1913—the day that the eight-hour day became Colorado law, named after the then-president of the U.M.W.A.

KERF. An undercut below a vein of coal.

LARD OIL. or *hog oil*—early oil used with string wick in lamp that hung on miner's cap when he worked underground.

LIGNITE. A lesser-quality coal found at the western edge of the northern field.

MEASURE. Rocks that have something in common geologically.

MOTORMAN. Worker who used electric machines underground.

NARROW WORK. The enlarging of tunnels for coal car passage—considered dead work by the operator, hence, not paid for.

NIPPER. A youngster who helped with the mules; later, he helped the motorman.

OPERATOR. Term for coal company management.

PEA COAL. A size of coal.

REDNECK. The operators' term for the striking miner, more often in the southern field, who traditionally wore a red kerchief around his neck. In case of trouble, the striker could be easily distinguished from the scab.

RUSH THE GROWLER. A man's first beer of the day.

SACCO-VANZETTI STRIKE. A test walkout held before the strike of 1927-1928 to determine if conditions were right for a strike. The New Jersey radicals, Nicola Sacco and Bartelomeo Vanzetti, were about to be executed for their alleged crimes.

SCAB. A laborer hired by the operator to take the place of a striking miner.

SCRIP. Paper issued by operators for wages instead of money, redeemable only at company stores. Sometimes called *flicker*.

SCREENING. The process of sorting coal by size. Early coal operators wanted miners to screen the coal with no additional wage.

SEAM. A coal deposit.

SHOT FIRER. A worker who specialized in setting explosive devices to free the coal.

SLACK. The poorest grade of coal, which could be used in some industries to produce steam.

SLAG. The coal that has disintegrated from chunks to powder.

SLOPE. Type of coal mine where the seams were close to the ground's surface, reached by a sloping tunnel.

SPOTTER. Company spy.

SPRAG. A twelve- to fourteen-inch iron or wooden piece with a crossbar at one end that was inserted between the wheel spokes of a coal car to slow down its progress, particularly along a sloped passageway.

SQUIB. An exploding device that, when lit, flew down a narrow hole drilled into a vein of coal.

SUB-BITUMINOUS. A type of coal found in the northern field that was lower in quality than the harder anthracite.

SUBSIDENCE. The sinking of unstable ground over empty coal rooms, which, in turn, causes damage to buildings and road.

TIN-WILLY. A nickname for members of the Colorado militia.

TIPPLE. A wooden tower or, later, a building over the mine shaft that brought loaded coal cars to the surface by means of a pulley and hoist in order to tip the loaded car into a waiting boxcar.

TRAPPER BOY. A young man hired to open and close tram doors in the mine.

TRIP. The pulling of coal cars.

WAGON MINE. Early simple coal mine where the coal was loaded onto a wagon for market.

WEIGH BOSS. A company man who weighed each miner's tonnage, sometimes to the company's advantage.

WHITE COAL. Miners' term for electricity.

WHITE DAMP. Carbon monoxide gas, hard to detect, found in the mines.

WHITE GOLD. Silver.

WHITE MULE. Bootleg whiskey.

WOBBLIE. A member of the Industrial Workers of the World, a labor movement that provided leadership for the strike of 1927-1928.

YELLOW DOG CONTRACT. A management requirement that an employee promise in writing not to join a union.

APPENDIX B

GEOLOGY OF THE NORTHERN COAL FIELD

The coal outcroppings in the Northern Coal Field were the result of a Lower Laramie geologic formation that thrust up in earlier ages as the mountains and foothills of Colorado pushed up in the same manner.

The veins of coal, some as wide as fourteen feet, were laid down one hundred million years ago during the Cretaceous era by a process of metamorphism, a change in the structure, texture, and composition of a layer associated with mountain building, when the Boulder valley was covered by a series of inland seas and swamps. As the humus and debris built up from the swampy environment, and as other deposits overlaid and exerted pressure, coal measures formed. From every fourteen feet of swamp debris, one foot of coal could develop.

The eastern third of Boulder County was underlain with coal deposits, but none of them were of high quality. The coal beds near the mountains produced a brown lignite that was soft, volatile, and low grade in character. The rest of the northern field produced a sub-bituminous coal, low grade in quality as well, but somewhat harder than the lignite. The hardest of coals, anthracite, was not found in the northern field, but was mined in Gunnison and Garfield counties, elsewhere in Colorado. Most anthracite is an older form of coal, laid down 280 to 345 million years ago during the Carboniferous era.

The coal beds were deeper at the northern end of the field; coal seams as thick as sixty-five feet have been found. Most of them lay parallel to the land surface except for the areas closest to the mountains. Those lignite beds tilted toward the mountains, a condition that allowed for slope mines rather than shafts, which could be as deep as three hundred feet.

Toward the southern end of the field were deposits of iron ore, hematite associated with sandstone layers, a combination that excited Joseph Marshall in the 1860s.

Throughout the Northern Coal Field are occasional oil and gas wells, a combination that is not unusual in association with a coal field.

NOTES

INTRODUCTION

1. Verse and chorus by Merle Travis, "Dark as a Dungeon" (Hollywood: American Music Inc., 1947).
2. W.B. Vickers, "History of Colorado" in *History of Clear Creek and Boulder Valleys* (Chicago: O.L. Baskin and Company, 1880), p. 18.
3. Ibid.
4. Philip S. Foner, ed., *Mother Jones Speaks: Collected Speeches and Writings* (New York: Monad Press, 1983), p. 106.
5. Wilbur Fiske Stone, *History of Colorado* (Chicago: S.J. Clark Publishing Company, 1918-1919), p. 449.
6. Quoted in Barron B. Beshoar, *Out of the Depths: the Story of John R. Lawson, a Labor Leader* (Denver: Colorado Labor Historical Committee of the Denver Trades and Labor Assembly, 1942), p. 209.

CHAPTER ONE

1. W.B. Vickers, "History of Colorado" in *History of Clear Creek and Boulder Valleys* (Chicago: O.L. Baskin and Company, 1880), p. 29.
2. Robert G. Athearn, *High Country Empire: the High Plains and Rockies* (Lincoln: University of Nebraska Press, 1960), p. 4.
3. H.N. Eavenson, *The First Century and a Quarter of the American Coal Industry* (Pittsburg: Koppers Building, 1942), p. 4.
4. Arthur Lake, *Geology of Colorado Coal Deposits*, Report for the State School of Mines, Golden, delivered October 31, 1889, p. 29.
5. *Rocky Mountain News*, September 29, 1859.
6. Ibid., September 10, 1859.
7. Nolie Mumey, *Life of Jim Baker* (New York: Interland Publishing Company, 1972), pp. 156-157.
8. Arthur Lake, *Geology*, p. 29.
9. *Boulder County News*, November 14, 1873.
10. Court House Collections, Box 13, Carnegie Branch Library for Local History, Boulder, Colorado.
11. Jennie Stewart, *Boulder County Pioneers* (Boulder: Arapahoe D.A.R., 1946-1948), p. 23.
12. Albert B. Sanford, "The First Iron Manufactory in Colorado," *Colorado Magazine*, vol. 8 (1931), p. 146.
13. Bayard Taylor, *Colorado: A Summer Trip* (New York: Putnam, 1867), p. 41.
14. Ibid., p. 157.
15. Ibid.
16. Samuel Bowles, *Our New West* (Hartford: Hartford Publishing Company, 1869), p. 195.

CHAPTER TWO

1. Amos Bixby, "History of Boulder County" in *History of Clear Creek and Boulder Valleys* (Chicago: O.L. Baskin and Company, 1880), p. 395.

2. Frank Hall, *History of Colorado* (Chicago: The Blakeley Printing Company, 1890), vol. 1, p. 474.
3. David Boyd, *A History: Greeley and the Union Colony of Colorado* (Greeley: The Greeley Tribune Press, 1890), p. 65.
4. Interview with coal miner Basil Beverstock, July 16, 1937, as quoted in Hazel Alice Glenny, "A History of Labor Disputes in the Northern Colorado Coal Mining Fields with Emphasis on the 1927-1928 Strike" (M.A. thesis, University of Colorado at Boulder, 1938).
5. Eric Margolis, "Western Coal Mining as a Way of Life," *Journal of the West*, vol. 24, no. 3 (July 1985), p. 14.
6. Ibid.
7. *Boulder County News*, March 8, 1872.
8. Ibid.
9. *Erie, Yesterday and Today* (Erie High School: 1967), p. 17.
10. Ibid.
11. Isabella Bird, *A Lady's Life in the Rocky Mountains* (Norman: University of Oklahoma Press, 1960), p. 197, first published by John Murray in London, 1879.
12. Dennis S. Grogan, "Unionization in Boulder and Weld Counties to 1890," *Colorado Magazine*, vol. 54, no. 4 (1967), p. 327.
13. Ibid., p. 328.
14. *Boulder County News*, April 20, 1877.
15. Jane Barker, "Over the Shoulder," Boulder *Daily Camera*, May 16, 1971.
16. Ibid.
17. *Greeley Sun*, May 31, 1876.
18. Ibid.

CHAPTER THREE

1. *Boulder News and Courier*, September 26, 1879.
2. Dee Brown, *Hear That Lonesome Whistle Blow* (New York: Holt, Rinehart and Winston, 1977), p. 184.
3. Amos Bixby, "History of Boulder County" in *History of Clear Creek and Boulder Valleys* (Chicago: O.L. Baskin and Company, 1880), p. 421.
4. Mel Griffiths, "Those Marvelous Mountain Mules," *Empire Magazine, Denver Post,* March 27, 1955, pp. 8-9.
5. Interview with William McShane by H. Lee Scamehorn, Jan-July 1968, quoted in James Whiteside, "Protecting the Life and Limb of Our Workmen: Coal Mining Regulation in Colorado, 1883-1920" (Ph.D. diss., University of Colorado at Boulder, 1986), p. 74.
6. Eleanor H. Ayer, *Green Light on the Tipple: Pioneering the Carbon Valley* (Frederick, Colorado: The Platte N Press, 1978), p. 13.
7. Duncan Emrich, collector, *Casey Jones and Other Ballads of the West* (n.p., 1942), p. 16.
8. Ibid.
9. Interview with Thomas Allen by Marjorie Harkness, March 9, 1962, p. 8, Colorado Division of State Archives and Public Records, Department of Natural Resources, Division of Mines Administration, File 1883-1972, Box 43919.
10. Ibid., p. 9.
11. *Greeley Tribune*, May 26, 1970, p. 24.
12. *Boulder County News*, May 31, 1878.
13. *Boulder County Herald*, January 20, 1886.

CHAPTER FOUR

1. Amos Bixby, "History of Boulder County" in *History of Clear Creek and Boulder Valleys* (Chicago: O.L. Baskin and Company, 1880), p. 421.
2. *United States Census*, 1880.
3. *Boulder News and Courier*, May 6, 1881.
4. Donald J. McClurg, "Labor Organization in the Coal Mines of Colorado, 1878-1933" (Ph.D. diss., University of California at Berkeley, 1959), p. 42.
5. Ibid.
6. *First Annual Report of the State Inspector of Coal Mines, 1884* (Denver: 1884), pp. 4-5.
7. Ibid., p. 9.
8. Carlos Schwantes, "The Concept of the Wageworkers' Frontier; a Framework for Future Research," *Western Historical Quarterly*, vol. 18, no. 1 (January 1987), p. 41.
9. Interview with Earl Stucker, Delta, May 8, 1978, quoted in Erick Margolis, "Western Coal Mining as a Way of Life," *Journal of the West*, vol. 24, no. 3 (July 1985), p. 14.
10. *First Biennial Report of the Bureau of Labor Statistics, 1887-1888,* (Denver: 1888), p. 59.
11. Ibid., p. 292.
12. Ibid., p. 82.

CHAPTER FIVE

1. *Third Biennial Report of the Bureau of Labor Statistics, 1891-1892,* (Colorado Springs: 1892), p. 167.
2. *Rocky Mountain News*, November 8, 1893.
3. Much of this material is from the work of Carla Joan Atchison, "Nativism in Colorado Politics" (M.A. thesis, University of Colorado at Boulder, 1977).
4. Donald J. McClurg, "Labor Organization in the Coal Mines of Colorado, 1878-1933" (Ph.D. diss., University of California at Berkeley, 1959).
5. Interview with Kenneth Brown, oral history tape, Lafayette Public Library.
6. Allen Kent Powell, *The Next Time We Strike* (Logan: Utah State University Press, 1985), p. 86.
7. *Second Biennial Report of the Bureau of Labor Statistics, 1889-1890,* (Denver: 1890), p. 38.
8. Ibid., p. 39.
9. Ibid.
10. *Third Biennial Report of the Bureau of Labor Statistics, 1891-1892,* (Colorado Springs: 1892), p. 176.
11. Ibid.
12. *Denver Republican*, October 25, 1893.
13. *Session Laws of Colorado*, 1897 (Denver: 1897), p. 137.
14. *Fifth Biennial Report of the State Inspector of Coal Mines, 1901-1902*, Henry Denman (Denver: 1902), p. 6.

CHAPTER SIX

1. Carolyn Conarroe, *The Louisville Story* (Louisville, Colorado: Louisville Times, Inc., 1978), p. 34.
2. *Rocky Mountain News*, February 9, 1901.
3. Ibid.
4. *Black Diamond World*, November 21, 1903.
5. Quoted in an article by John Andrews, Boulder *Daily Camera*, January 25, 1976.

6. Ibid.
7. *Denver Times*, July 17, 1902.
8. Eleanor H. Ayer, *Green Light on the Tipple: Pioneering the Carbon Valley* (Frederick, Colorado: Platte N Press, 1978), p. 3.

CHAPTER SEVEN

1. Donald J. McClurg, "Labor Organization in the Coal Mines of Colorado, 1878-1933" (Ph. D. diss., University of California at Berkeley, 1959).
2. *Eighth Biennial Report of the Bureau of Labor Statistics, 1901-1902*, (Denver: 1902), p. 137.
3. W.T. Davis, "The Strike War in Colorado," *Outlook*, April 22, 1914, pp. 67-73.
4. Quoted in Sylvia Ruland, "Mother Jones: the Most Dangerous Woman in America," *Empire Magazine, Denver Post*, June 7, 1981, p. 49.
5. Priscilla Long, *Mother Jones, Woman Organizer* (Cambridge, Mass.: Red Sun Press, 1976), p. 2.
6. Sylvia Ruland, p. 49.
7. Priscilla Long, p. 33.
8. Philip Foner, ed., *Mother Jones Speaks: Collected Speeches and Writings* (New York: Monad Press, 1983), pp. 105-106.
9. Ibid., p. 106.
10. Ibid., pp. 106-107. (Also printed in the *Denver Post*, November 22, 1903.)
11. George S. McGovern and Leonard F. Guttridge, *The Great Coalfield War* (Boston: Houghton Mifflin Company, 1972), p. 75.
12. Donald J. McClurg, p. 141.
13. Quoted in Marjorie Hornbein, "Josephine Roche: Social Worker and Coal Operator," *Colorado Magazine*, vol. 53, no. 3 (1976), p. 244.

CHAPTER EIGHT

1. George S. McGovern and Leonard F. Guttridge, *The Great Coalfield War* (Boston: Houghton Mifflin Company, 1972), pp. 73-74.
2. "Private War in Colorado," *Outlook*, vol. 107 (May 9, 1914), p. 61, as quoted in Maxine Benson, "Labor and the Law in Colorado: People vs. John Lawson" (M.A. thesis, University of Colorado at Boulder, 1962).
3. *Twelfth Biennial Report of the Bureau of Labor Statistics, 1909-1910* (Denver: 1910), p. 11.
4. W.T. Davis, "The Southern Colorado Coal Strike," *Outlook*, vol. 106 (1914), pp. 24-26.
5. Boulder *Daily Camera*, March 7, 1910.
6. Ibid.
7. Barron B. Beshoar, *Out of the Depths: the Story of John R. Lawson, a Labor Leader* (Denver: Colorado Labor Historical Committee of the Denver Trades and Labor Assembly, 1942), p. 35.
8. *Twelfth Biennial Report*, p. 266.
9. Ibid., p. 267.
10. Ibid., pp. 267-268.
11. Boulder *Daily Camera*, August 17, 1910.
12. Donald J. McClurg, "Labor Organization in the Coal Mines of Colorado, 1878-1933," (Ph.D. diss., University of California at Berkeley, 1959), p. 153.
13. Boulder *Daily Camera*, August 17, 1910.
14. Ibid., August 29, 1910.

15. Ibid., September 19, 1910.
16. Ibid.
17. *Rocky Mountain News*, December 1, 1910, p. 6.
18. Ibid., December 24, 1910.
19. Barron B. Beshoar, p. 26.
20. *Rocky Mountain News*, February 3, 1911.
21. Boulder *Daily Camera*, February 24, 1911.
22. *Rocky Mountain News*, April 17, 1911.
23. Barron B. Beshoar, p. 28.
24. *Rocky Mountain News*, July 6, 1911.
25. *Thirteenth Biennial Report of the Bureau of Labor Statistics, 1911-1912* (Denver: 1912), p. 138.
26. R. Michael Collard, "The Dude Searcy Affair," paper, 1976, Louisville Public Library, Louisville, Colorado.
27. Ibid.
28. *Thirteenth Biennial Report*, p. 139.

CHAPTER NINE

1. Leroy Hafen, ed., *Colorado and its People*, vol. 2 (New York: Lewis Historical Publishing Company, Inc., 1948), p. 322.
2. Quoted in Zeese Papanikolas, *Buried Unsung: Louis Tikas and the Ludlow Massacre* (Salt Lake City: University of Utah Press, 1982), p. 171.
3. "Statement of the Strike Situation in Colorado," Junior Order of American Mechanics, Kensington Council No. 6 (Denver: C.W. Varnum, 1914), pp. 10-16, as quoted in M. James Kedro, "Czechs and Slovaks in Colorado, 1860-1920," *Colorado Magazine* vol. 54, no. 2 (Spring 1977), p. 93.
4. Ibid.
5. Zeese Papanikolas, p. 51.
6. Ibid., pp. 51-52.
7. Ibid., p. 52.
8. Boulder *Daily Camera*, April 22, 1914.
9. *Denver Express*, October 7, 1914.
10. Donald J. McClurg, "Labor Organization in the Coal Mines of Colorado, 1878-1933" (Ph.D. diss., University of California at Berkeley, 1959), p. 186.
11. Ben Lindsey, undated article, Josephine Roche papers, Western Historical Collections, University of Colorado at Boulder.
12. *Rocky Mountain News*, April 20, 1975.
13. Ibid., May 5, 1914.
14. *Denver Express*, September 24, 1913.
15. Zeese Papanikolas, interview with Mary Thomas O'Neal, p. 159.
16. Letter of Louis Tikas, February 10, 1914, United Mine Workers of America Records, District 15 Correspondence.
17. Ibid.
18. *The Military Occupation of the Coal Strike Zone of Colorado by the Colorado National Guard, 1913-1914* (Denver: 1914), pp. 97098, as quoted in McClurg dissertation, p. 244.
19. Elias Ammons, "The Colorado Coal Strike," *North American Review*, vol. 200 (July 1914), pp. 35-44.
20. Walter Fink, *The Ludlow Massacre* (Denver: District 15, U.M.W.A., 1914), p. 9.
21. Ibid.
22. *Hearings Before the United States Commission on Industrial Relations*, Sen. Doc., 64th Cong., lst sess. (Washington D.C.: 1916), p. 8430.

23. Barron B. Beshoar, *Out of the Depths, the Story of John R. Lawson, a Labor Leader* (Denver: Colorado Labor Historical Committee, Denver Trades and Labor Assembly, 1942), p. 185.

CHAPTER TEN

1. *Denver Express*, April 22, 1914.
2. Deposition by Sanford Buster, Sheriff of Boulder County, as quoted in the *Louisville Times*, Centennial Edition, August 17, 1978, p. 10.
3. Ibid.
4. Ibid.
5. Ibid., p. 11.
6. Ibid., pp. 11-12.
7. Ibid., p. 12.
8. Carolyn Conarroe, *The Louisville Story* (Louisville, Colorado: Louisville Times, Inc., 1978), p. 26.
9. Barron B. Beshoar, *Out of the Depths: the Story of John R. Lawson, a Labor Leader* (Denver: Colorado Labor History Committee, Denver Trades and Labor Assembly, 1942), p. 185.
10. George S. McGovern and Leonard F. Guttridge, *The Great Coalfield War* (Boston: Houghton Mifflin Company, 1972), p. 344.

CHAPTER ELEVEN

1. Boulder *Daily Camera*, March 30, 1915.
2. Interview with Daisy Bracegirdle Swallow, 1987.
3. Ibid.
4. Boulder *Daily Camera*, April 6, 1915.
5. Ibid., April 9, 1915.
6. Ibid., April 17, 1915.
7. Interview with Mike Livoda, August 10, 1953, as quoted in Donald J. McClurg, "Labor Organization in the Coal Mines of Colorado, 1878-1933" (Ph.D. diss., University of California at Berkeley, 1959), p. 432.
8. Boulder *Daily Camera*, June 4, 1921.
9. Quoted in Robert Alan Goldberg, *The Hooded Empire: the KKK in Colorado, 1921-1932* (Urbana, Illinois: University of Illinois Press, 1981), p. 7.
10. *Boulder News-Herald*, August 6, 1924.
11. *Rocky Mountain American*, April 24, 1925.
12. Ibid.
13. James H. Davis, "Colorado Under the Klan," *Colorado Magazine*, vol. 42, no. 2 (1965), pp. 93-108.
14. *Rocky Mountain American*, February 20, 1925.
15. Ibid.
16. Eleanor H. Ayer, *Green Light on the Tipple: Pioneering the Carbon Valley* (Frederick, Colorado: The Platte N Press, 1978), p. 82.
17. Ibid.
18. Quoted in Robert G. Athearn, *The Coloradans* (Albuquerque: University of New Mexico Press, 1976), p. 246.
19. *Rocky Mountain American*, July 23, 1925.

CHAPTER TWELVE

1. Quoted in Melvyn Dubofsky, *We Shall Be All: a History of the I.W.W.* (Chicago: Quadrangle Books, 1969), front paper.
2. Donald J. McClurg, "The Colorado Coal Strike of 1927—Tactical Leadership of the I.W.W.," *Labor History*, vol. 4 (1963), p. 90.
3. Brooklyn *Eagle*, Brooklyn, New York, November 24, 1927, p. 8.
4. "The Colorado Coal Battle," *Outlook*, vol. 146 (December 7, 1927), p. 422.
5. Colorado Industrial Commission, *Tenth Report for the Biennium December 1, 1926 to November 30, 1928* (Denver: 1928), p. 61.
6. Boulder *Daily Camera* December 17, 1927, p. 2.
7. "Bloody Monday Again in Colorado," *Independent*, vol. 129 (December 31, 1927), p. 656.
8. Interview with Jessie Irwin James.
9. John S. Gambs, *Decline of the I.W.W.* (New York: Columbia University Press, 1932), p. 145.
10. Interview with A.S. Embree by Donald J. McClurg, "Labor Organization in the Coal Mines of Colorado, 1878-1933" (Ph.D. diss., University of California at Berkeley, 1959), fn., p. 445.
11. Hazel Alice Glenny, "A History of Labor Disputes in the Northern Colorado Coal Mining Fields with Emphasis on the 1927-1928 Strike" (M.A. thesis, University of Colorado at Boulder, 1938), p. 7.
12. Charles J. Bayard, "The 1927-1928 Colorado Coal Strike," *Pacific Historical Review*, vol. 32, no. 3 (August 1963), p. 237.
13. Donald J. McClurg, "Labor Organization in the Coal Mines," p. 469.
14. *Rocky Mountain News*, October 19, 1927.
15. Trinidad *Chronicle-News*, October 24, 1927.
16. A.S. Embree, *Industrial Solidarity*, No. 479, December 27, 1927.
17. Ibid.
18. *Denver Post,* October 27, 1927.
19. Leroy Hafen, *Colorado: the Story of a Western Commonwealth* (New York: AMS Press, 1970), p. 288.
20. Ben Lindsey and Harvey J. O'Higgins, *The Beast* (New York: Doubleday, Page and Company, 1910), p. 233.
21. *Rocky Mountain News*, October 27, 1927.
22. Perry Eberhart, "Trouble in Serene," *Denver Westerners Brandbook*, vol. 29 (1973), p. 205.
23. *Rocky Mountain News*, November 18, 1927.
24. Quoted in the Trinidad *Chronicle-News*, November 15, 1927.
25. *Rocky Mountain News*, November 12, 1927.
26. Interview with Jessie Irwin James, March 1987.
27. *Colorado Labor Advocate*, November 17, 1927, p. 3.
28. Josephine Roche, "Mines and Men," *Survey*, vol. 61 (December 15, 1928), p. 341.

CHAPTER THIRTEEN

1. Interview with Lawrence Amicarella, June 1987.
2. Colorado. Coroner of Weld County. *Inquest on the Bodies of John Eastenas and Nick Spanudaklis*, December 23, 1927.
3. *Colorado Labor Advocate*, November 24, 1927.
4. Interview with Mary Borstnick, 1987.
5. *Rocky Mountain News*, November 22, 1927, p. 4.

6. Ibid., November 24, 1927.
7. *Nation*, January 11, 1928.
8. *Rocky Mountain News*, November 24, 1927.
9. Colorado Industrial Commission, *Tenth Report for the Biennium, December 1, 1926 to November 30, 1928,* pp. 49-62.
10. *Rocky Mountain News*, December 18, 1927.
11. Ibid.
12. Ibid.
13. Ibid.
14. Ibid.
15. Ibid.
16. McQuarrie letters, Box 7, Josephine Roche papers, Western Historical Collections, University of Colorado at Boulder.
17. Ibid.
18. *Rocky Mountain News*, February 20, 1928.
19. Donald J. McClurg, "The Colorado Coal Strike of 1927—Tactical Leadership of the I.W.W.," *Labor History*, vol. 4 (1963), p. 92.
20. Donald J. McClurg, "Labor Organization in the Coal Mines of Colorado, 1878-1933" (Ph.D. diss., University of California at Berkeley, 1959), p. 532.
21. Melvyn Dubofsky, *We Shall Be All: a History of the I.W.W.* (Chicago: Quadrangle Books, 1969), p. 484.
22. Donald J. McClurg, "Labor Organization in the Coal Mines,' p. 521.
23. Hazel Alice Glenny, "A History of Labor Disputes in the Northern Colorado Coal Mines with emphasis on the 1927-1928 Strike" (M.A. thesis, University of Colorado at Boulder: 1938), pp. 114-125.
24. Ibid.
25. John S. Gambs, *25 Years of Industrial Unionism* (n.p.,n.d.), p. 58, as quoted in the *Decline of the I.W.W.* (New York: 1932), p. 153.

CHAPTER FOURTEEN

1. Mary Van Kleeck, *Miners and Management* (New York: Russell Sage Foundation, 1934), p. 248.
2. Ibid., p. 45.
3. Boulder *Daily Camera*, August 14, 1934.
4. Ibid.
5. Ibid., July 12, 1935.
6. Interview, Anna Barrowman, oral history tape, Lafayette Public Library.
7. Mary Van Kleeck, p. 112.
8. *Monthly Labor Review*, August 1932.
9. Robert G. Athearn, *The Coloradans* (Albuquerque: University of New Mexico Press, 1976), p. 274.
10. Interview with Lawrence Amicarella, 1987.
11. *Chicago Sunday Tribune*, September 29, 1940.
12. Interview, Coal Project, 1976, University of Colorado at Boulder.

CHAPTER FIFTEEN

1. Marjorie Hornbein, "Josephine Roche: Social Worker and Coal Operator," *Colorado Magazine*, vol. 53, no. 3 (1976), p. 259.
2. Kurt Chandler, "The Black Lung Nightmare: The Fight for Benefits Rages On," Boulder *Daily Camera*, November 14, 1984.

3. Glen Troelstrug, "Bureaucracy Stifles Black-Lung Victims' Aid Hope," *Denver Post*, July 3, 1977.

4. Ibid.

5. *Lafayette Leader*, December 24, 1970.

6. Ibid.

7. Boulder *Daily Camera*, December 5, 1978.

8. *Longmont Times-Call*, January 18, 1987.

9. Boulder *Daily Camera*, October 12, 1975.

10. Ibid.

11. John Craggs, "Marshall Once Was Scene of Boulder's Flourishing Coal Mining Industry," *Focus*, Boulder *Daily Camera*, January 25, 1970.

BIBLIOGRAPHY

Books and Articles from Scholarly Journals

Abbot, Carl. *Colorado, a History of the Centennial State*. Boulder: Colorado Associated University Press, 1974.

Ammons, Elias. "The Colorado Coal Strike." *North American Review*, vol. 200 (July 1914), pp. 35-44.

Athearn, Robert G. *The Coloradans*. Albuquerque: University of New Mexico Press, 1976.

———. *High Country Empire: the High Plains and Rockies*. Lincoln: University of Nebraska Press, 1960.

Atkins, James. *Human Relations in Colorado: an Historical Record*. Denver: Department of Education, State of Colorado, October 1968.

Atkinson, Rev. A.A. *The Church and Industrial Warfare: a Report on the Labor Trouble in Colorado and Michigan*. Federal Council of Churches of Christ of America, June 10, 1914.

Atkinson, Linda. *Mother Jones: the Most Dangerous Woman in America*. New York: Crown, 1978.

Ayer, Eleanor H. *Green Light on the Tipple: Pioneering the Carbon Valley*. Frederick, Colorado: The Platte N Press, 1978.

Bancroft, Hubert Howe. "History of Nevada, Colorado and Wyoming, 1540-1888." In *Works*, vol. 25. San Francisco: The History Company, 1890.

Bayard, Charles J. "The 1927-1928 Colorado Coal Strike." *Pacific Historical Review*, vol. 32, no. 3 (August 1963), pp. 235-251.

Beshoar, Barron B. *Out of the Depths: the Story of John R. Lawson, a Labor Leader*. Denver: Colorado Labor Historical Committee of the Denver Trades and Labor Assembly, 1942.

Bird, Isabella. *A Lady's Life in the Rocky Mountains*. Norman: University of Oklahoma Press, 1960. First published by John Murray in London, 1879.

Bird, Stewart, Dan Georgakas, Deborah Shaffer. *Solidarity Forever: An Oral History of the I.W.W.* Chicago: Lake View Press, 1985.

Bixby, Amos. "History of Boulder County" in *History of Clear Creek and Boulder Valleys*. Chicago: O.L. Baskin and Company, 1880.

Bowles, Samuel. *Our New West*. Hartford: Hartford Publishing Company, 1869.

Boyd, David. *A History of Greeley and the Union Colony of Colorado*. Greeley: Greeley Tribune, 1890.

Brown, Dee. *Hear That Lonesome Whistle Blow*. New York: Holt, Rinehart and Winston, 1977.

Buchanan, Joseph. *The Story of a Labor Agitator*. New York: The Outlook Company, 1903.

Chenery, William. *So it Seemed*. New York: Harcourt Brace and Company, 1952.

Committee of Coal Mine Managers. *Facts Concerning the Struggle in Colorado for Industrial Freedom*. Series I, September 21, 1914.

Conarroe, Carolyn. *The Louisville Story*. Louisville: Louisville Times, Inc., 1978.

Creel, George. *Rebel at Large: Recollections of 50 Crowded Years*. New York: G. P. Putnam, 1947.

Dickerson, Oliver M. "The Labor Movement" in *Colorado and its People*. 4 vols. Leroy Hafen, ed. New York: Lewis Historical Publishing Company, Inc., 1948.

Dubofsky, Melvyn. *We Shall Be All: a History of the I. W. W.* Chicago: Quadrangle Books, 1969.

Erie, Yesterday and Today. Erie High School, 1967. Karen Adelfang, adviser.

Fetherling, Dale. *Mother Jones, the Miners' Angel*. Carbondale and Edwardsville: Southern Illinois University Press, 1974.

Fink, Walter. *The Ludlow Massacre*. Denver: District 15, U.M.W.A., 1914.

Foner, Philip S. *American Labor Songs of the Nineteenth Century*. Urbana: University of Illinois Press, 1975.

———, ed. *Mother Jones Speaks: Collected Speeches and Writings*. New York: Monad Press, 1983.

Fossett, Frank. *Colorado*. New York: C.G. Crawford, 1879.

Fritz, Percy S. *Colorado, the Centennial State*. New York: Prentice-Hall, Inc., 1941.

Gambs, John S. *The Decline of the I. W. W.* New York: Columbia University Press, 1932.

Geffs, Mary. *Under Ten Flags, a History of Weld County*. Greeley: McVey Printery, 1938.

Goldberg, Robert Alan. *The Hooded Empire: the KKK in Colorado, 1921-1932*. Urbana: University of Illinois Press, 1981.

Greiner, Jean M. "The Golden, Boulder and Caribou." *Colorado Magazine*, vol. 44, no. 2 (1967), pp. 307-333.

Grogan, Dennis S. "Unionization in Boulder and Weld Counties to 1890." *Colorado Magazine*, vol. 44, no. 4, (1967), pp. 324-341.

Hall, Frank. *History of Colorado*. 3 vols. Chicago: The Blakeley Printing Company, 1890 and 1891.

Henderson, Charles W. "Mining in Colorado" in *History of Colorado*. James H. Baker and Leroy Hafen, eds. Denver: Linderman Company, 1927.

Hollister, Ovando James. *The Mines of Colorado*. Springfield, Mass.: Samuel Bowles and Company, 1867.

Hornbein, Marjorie. "Josephine Roche: Social Worker and Coal Operator." *Colorado Magazine*, vol. 53, no. 3 (1976), pp. 243-260.

Kauffman, Mary Ruth. *Sarah, Her 100 Years*. Boulder: Gambill Properties, 1982.

Kedro, M. James. "Czechs and Slovaks in Colorado, 1860-1920." *Colorado Magazine*, vol. 54, no. 2 (Spring 1977), p. 93.

Knight, Harold V. *Working in Colorado: a Brief History of the Colorado Labor Movement*. Denver: World Press, 1971.

Kutes, Andrew. *History of the Czechs and Slovaks in the State of Colorado, 1876-1976*. Denver: 1976.

"Langford, Augustine Gallet, 1834-1885." *Colorado Magazine*, vol. 22, (1945), pp. 213-215.

Lindsey, Benjamin B. and Harvey J. O'Higgins. *The Beast*. New York: Doubleday, Page and Company, 1910.

Long, Priscilla. *Mother Jones, Woman Organizer*. Cambridge, Mass.: Red Sun Press, 1976.

Margolis, Eric. "Western Coal Mining as a Way of Life: an Oral History of the Colorado Coal Miners to 1914." *Journal of the West*, vol. 24, no. 3 (July 1985), pp. 5-115.

McClurg, Donald J. "The Colorado Coal Strike of 1927—Tactical Leadership of the I.W.W." *Labor History*, vol. 4 (1963), pp. 68-92.

McGovern, George S. and Leonard F. Guttridge. *The Great Coalfield War*. Boston: Houghton Mifflin Company, 1972.

McMahan, Ronald L. "Rang-U-Tang: the I.W.W. and the 1927 Colorado Coal Strike" in *At the Point of Production: the Local History of the I.W.W.* Joseph R. Conlin, ed. Westport, Conn.: Greenwood Press, 1981.

Mitick, Ellsworth C. "A History of Mining Manufacture in Colorado." *Colorado Magazine*, vol. 24 (November 1947), p. 226.

Mumie, Nolie. *Life of Jim Baker*. Denver: World Press, 1931.

O'Neal, Mary Thomas. *Those Damn Foreigners*. Hollywood: Minerva Printing and Publishing Company, 1971.

"Origin of Communities in Boulder County." *Boulder Genealogical Society Quarterly*, August 1974.

Papanikolas, Zeese. *Buried Unsung: Louis Tikas and the Ludlow Massacre*. Salt Lake City: University of Utah Press, 1982.

Powell, Allan Kent. *The Next Time We Strike*. Logan, Utah: Utah State University Press, 1985.

Sanford, Albert B. "The First Iron Manufactory in Colorado." *Colorado Magazine*, vol. 8 (1931), pp. 142-146.

Schoolland, John B. *Boulder in Perspective: From the Search of Gold to the Gold of Research*. Boulder: 1980.

Schwantes, Carlos. "The Concept of the Wageworkers' Frontier: a Framework for Future Research." *Western Historical Quarterly*, vol. 28, no. 1 (January 1987), pp. 39-55.

Selekman, Benjamin and Mary Van Kleeck. *Employee Representation in Coal Mines*. New York: Russell Sage Foundation, 1924.

Sinclair, Upton. *King Coal*. New York: 1917.

———. *The Coal War, a Sequel to King Coal*. Introduction by John Graham. Boulder: Colorado Associated University Press, 1976.

Smedley, Agnes. *Daughters of Earth*. Old Westbury, New York: Feminist Press, 1963.

Smith, Phyllis. *A Look at Boulder: From Settlement to City*. Boulder: Pruett Publishing Company, 1981.

Sprague, Marshall. *Colorado: a Bicentennial History*. New York: W.W. Norton, 1976.

Stewart, Jennie. *Boulder County Pioneers*. Boulder: Arapahoe D.A.R., 1946-1948.

Stone, Wilbur Fiske. *History of Colorado*. 4 vols. Chicago: S.J. Clark Publishing Company, 1918-1919.

Suffern, Arthur E. *Coal Miners' Struggle for Industrial Status*. New York: Macmillan, 1926.

Suggs, George G., Jr. "The Colorado Coal Miners' Strike, 1903-1904: a Prelude to Ludlow?" *Journal of the West*, vol. 12, no. 1.

Superior, a Folk History. Superior Community Involvement Program, 1983.

Taylor, Bayard. *Colorado: a Summer Trip*. New York: Putnam, 1867.

Tice, J.H. *Over the Plains and on the Mountains*. St. Louis: Industrial Age Printing, 1872.

Ubbelohde, Carl, Maxine Benson, and Duane A. Smith. *A Colorado History*. Rev. Boulder: Pruett Publishing Company, 1972.

Van Kleeck, Mary. *Miners and Management*. New York: Russell Sage Foundation, 1934.

Vickers, W.B. "History of Colorado." in *History of Clear Creek and Boulder Valleys.* Chicago: O.L. Baskin and Company, 1880.

Whiteside, James. "Protecting the Life and Limb of Our Workmen: Coal Mining Regulation in Colorado, 1883-1920." *Essays and Monographs in Colorado History.* Issue 4 (1986), pp. 2-24.

Collections

Costigan, Edward P. Papers. Western Historical Collections, University of Colorado at Boulder.

Doyle, Edward L. Papers. Western History Department, Denver Public Library, Denver, Colorado.

Fritz, Percy. Papers. Western Historical Collections, University of Colorado at Boulder.

Lawson, John R. Papers. Western History Department, Denver Public Library, Denver, Colorado.

Meyer, Erwin. Papers. Western Historical Collections, University of Colorado at Boulder.

Roche, Josephine. Papers. Western Historical Collections, University of Colorado at Boulder.

Rocky Mountain Fuel Company. Papers. Western Historical Collections, University of Colorado at Boulder.

———. Papers. Western History Department. Denver Public Library, Denver, Colorado.

Government Publications, Records, and Documents

Boulder County Records. Book 13, July 20, 1869, and Book 40, October 8, 1860. (Records for the short-lived Jackson County, Jefferson Territory.)

Colorado. Bureau of Labor Statistics. *First Biennial Report, 1887-1888.* James Rice and C.J. Driscoll. Denver: 1888.

———. *Second Biennial Report, 1889-1890.* Elmer F. Beckwith and James T. Smith. Denver: 1890.

———. *Third Biennial Report, 1891-1892.* E.J. Eaton and L. Bodine. Colorado Springs: 1892.

———. *Fourth Biennial Report, 1893-1894.* W.H. Montgomery and James Cowie. Denver: 1894.

———. *Seventh Biennial Report, 1889-1900.* Denver: 1900.

———. *Eighth Biennial Report, 1901-1902.* Denver: 1902.

———. *Ninth Biennial Report, 1903-1904.* Denver: 1904.

———. *Twelfth Biennial Report, 1909-1910.* James B. Pearce and Edwin V. Brake. Denver: 1911.

———. *Thirteenth Biennial Report, 1911-1912.* Denver: 1912.

Colorado. Weld County Coroner. *Inquest on the Bodies of John Eastenas and Nick Spanudaklis.* December 23, 1927.

Colorado. State Inspector of Coal Mines. *First Annual Report, 1884.* John McNeil. Denver: 1884.

———. *Fifth Biennial Report. 1891-1892.* John McNeil. Denver: 1892.

———. *Ninth Biennial Report, 1899-1900.* Henry Denman. Denver: 1900.

———. *Tenth Biennial Report, 1901-1902.* Henry Denman. Denver: 1902.

———. *Twelfth Biennial Report, 1905-1906.* John D. Jones. Denver: 1907.

Colorado. *Inactive Mine Reclamation Plan.* 2 vols.

Colorado. *Mining Laws.* 1921.

United States. Commission on Industrial Relations. "Report on the Colorado Coal Strike." George West. 1915.

United States Congress. House. *Conditions in the Coal Mines of Colorado.* Hearings before a Subcommittee of the Committee on Mines and Mining. 63rd Cong., 2d sess., 1914, pursuant to H. Res. 387. 2 vols.

Theses, Dissertations, and Reports

Amicarella, Effie. "Lafayette—Town's History." Paper. Lafayette Public Library, Lafayette, Colorado. December 8, 1981.

Atchison, Carla Joan. "Nativism in Colorado Politics." M.A. thesis, University of Colorado at Boulder, 1971.

Benson, Maxine. "The Colorado Coal Strike, 1913-1914." Paper. Western Historical Collections, University of Colorado at Boulder.

———. "Labor and the Law in Colorado: People vs. John Lawson." M.A. thesis, University of Colorado at Boulder, 1962.

Buster, Sanford, Sheriff of Boulder County. "Shooting at the Hecla Mine, 1914." Deposition. Rocky Mountain Fuel Company Papers, Western History Department, Denver Public Library, Denver, Colorado.

Collard, R. Michael. "Historical Report on Monarch #2 Mine Explosion and the Joe Jaramillo Memorial Stone." Report, 1976. Carnegie Branch Library for Local Research, Boulder, Colorado.

———. "Historical Report on the Superior Cemetery." Report, 1976. Louisville Public Library, Louisville, Colorado.

———. "The Dude Searcy Affair." Report, 1976. Louisville Public Library, Louisville, Colorado.

Glenny, Hazel Alice. "A History of Labor Disputes in the Northern Colorado Coal Mining Fields with Emphasis on the 1927-1928 Strike." M.A. thesis. University of Colorado at Boulder, 1938.

Hogg, Maureen E. "The Proving Ground." Paper, 1977. University of Northern Colorado, Greeley, Colorado.

Lawson, Harry O. "The Colorado Coal Strike of 1927-1928." M.A. thesis. University of Colorado at Boulder, 1950.

McClurg, Donald J. "Labor Organization in the Coal Mines of Colorado, 1878-1933." Ph.D. diss. University of California at Berkeley, 1959.

Mernitz, Scott. "The Impact of Coal Mining on Marshall, Colorado and Vicinity: an Historical Geography of Environmental Changes." M.A. thesis. University of Colorado at Boulder, 1971.

Peck, Maynard Aurelius. "Some Economic Aspects of the Coal Industry in Boulder County, Colorado." Ph.D. diss. University of Colorado at Boulder, 1947.

Shuler, Bobbalee. "Scab Labor at the Columbine Mine, 1927-1928." M.A. thesis. Colorado State University, 1986.

"Statement and Report of the Belle Monte Iron and Coal Company of Colorado." New York: Sackett and MacKay, 1866.

Whiteside, James. "Protecting the Life and Limb of Our Workmen: Coal Mining Regulation in Colorado, 1883-1920." Ph.D. diss. University of Colorado at Boulder, 1986.

Periodicals

Atkinson, Henry A. "Why the Miners Struck." *Harper's Weekly* (May 23, 1914), p. 3.

Baker, Ray Stannard. "Reign of Lawlessness, Anarchy and Despotism in Colorado." *McClure's Magazine*, vol. 23 (March 1904), p. 43.

Berle, A.A. "The Colorado Mine War." *Bibliotheca Sacra* (October 1914).

"Colorado Tries a New Strategy." *Christian Century*, (January 28, 1931), p. 125.

"The Blood-Spilling in Colorado." *Literary Digest*, vol. 95, no. 10 (December 3, 1927), pp. 1-3.

Bower, Carolyn, "Survivor Recalls 1936 Mine Disaster." *Focus*, Boulder *Sunday Camera* (January 17, 1965).

Craggs, John. "Marshall was Once Scene of Boulder's Flourishing Coal Mining Industry." *Focus*, Boulder *Sunday Camera* (January 25, 1970).

Creel, George. "The High Cost of Hate." *Everybody's Magazine* (June 1914), p. 755.

Davis, Shirley. Collected articles, *Longmont Times-Call*, (October 30-31, 1976 and March 5-6, 1977).

Davis, W.T. "The Southern Colorado Coal Strike." *Outlook*, vol. 106 (1914), pp. 24-26.

———. "The Strike War in Colorado." *Outlook*, vol. 107 (April 22, 1914), pp. 67-73.

DeMers, Tom. "The Great Colorado Coal Strike." *Sunday Magazine*, Boulder *Daily Camera* (April 22, 1984).

Eberhart, Perry. "Trouble in Serene." *Denver Westerners Brandbook*, vol. 29 (1973), pp. 203-211.

Embree, Adolphus Stewart. "The Story of the Great Colorado Coal Strike." *Industrial Solidarity*, no. 478 (December 21, 1927); no. 479 (December 27, 1927), no. 481 (January 11, 1928).

Fitch, John A. "Law and Order: the Issue in Colorado." *Survey*, vol. 33, no. 10 (December 5, 1914), pp. 241-258.

———. "When Peace Comes to Colorado." *Survey*, vol. 32 (May 16, 1914).

Griffiths, Mel. "Those Marvelous Mountain Mules." *Empire Magazine, Denver Post* (March 27, 1955), pp. 8-9.

"Lafayette." *Colorado Prospector*, vol. 6, no. 11 (November 1975).

Lawson, John R. "History of Industrial Struggles in Colorado Coal Mines." *Colorado Labor Advocate*, December 31, 1931.

"Louisville." *Colorado Prospector*, vol. 6, no. 10 (October 1975).

Marfarlane, Peter Clark. "The War in Colorado." *Collier's* (June 6, 1914).

Marble, Samuel W. "Justice and Coal." *Christian Century* (January 19, 1928).

Meyer, Erwin F. "Six Killed, Twenty Wounded: a Case Study of Industrial Conflict." *Survey*, vol. 59, no. 10 (February 15, 1928), pp. 644-646.

Palmer, Frank L. "Bloody Colorado." *Nation* (November 16, 1927).

———. "War in Colorado." *Nation* (December 7, 1927).

———. "Rebellion in Colorado." *Nation* (January 11, 1928).

———. "Solidarity in Colorado." *Nation* (February 1, 1928).

Rice, Millard Milburn. "Bloody Monday Again in Colorado." *Independent* (December 31, 1927).

"R.M.F." *Time* (July 10, 1931).

Robinson, Helen Ring. *Independent*, vol. 78 (May 11, 1914). pp. 245-247.

Ruland, Sylvia. "Mother Jones: the Most Dangerous Woman in America." *Empire Magazine, Denver Post* (June 7, 1981).

Stark, Louis. "A Woman Unravels an Industrial Knot." *New York Times Magazine* (February 7, 1932).

"Union Angel." *Business Week* (June 17, 1944).

United Mine Workers Journal (November 11, 19, 26, 1903).

Villard, Oswald Garrison. "Josephine Roche for Governor of Colorado." *Nation* (June 13, 1934).

Newspapers and Articles

Barker, Jane. "When Coal was King," Boulder *Daily Camera*, April 25, 1971.

———. "Pennsylvanian Founds Erie." Boulder *Daily Camera*, May 16, 1971.

———. "Erie Once Entangled in Erie War." Boulder *Daily Camera*, May 23, 1971.

Bates, Betsy. "Mother of Lafayette Earned Residents' Admiration," Boulder *Daily Camera*, November 20, 1980.

Boulder *Daily Camera*, Souvenir Industrial Edition, December 15, 1902.

———. "Insurrection in Colorado During the Years 1913-1914." Special undated edition.

Burleigh, Linda. "The 'Baby Railroad.' " *Longmont Times-Call*, November 23, 1981.

Butman, Beverly. "Injury, Sickness, Low Pay have been Coal Miners' Way of Life." Boulder *Daily Camera*, November 24, 1974.

Chandler, Kurt. "The Black Lung Nightmare: the Fight for Benefits Rages On." Boulder *Daily Camera*, November 14, 1984.

Conarroe, Carolyn. "A Heritage of Coal Production," *Louisville-Lafayette News*, February 10, 1977.

Cramer, Anne. "All About Mary." *Lafayette Times*, October 1, 1980.

Crossen, Forest. "Mrs. Ballard Recalls Early Days of Her Residence in Lafayette." Boulder *Daily Camera*, July 10, 1956.

Finholm, Valerie. "It's been a Hard Road from Mining to Marketing in Boulder County." Boulder *Daily Camera*, June 26, 1983.

Frizell, Ken and John Leach. "Coal Strip Mining Efforts Reaching Area Residents." Boulder *Daily Camera*, October 12, 1975.

Gaeddert, Beth. "Old Coal Fire Burning Under Boulder Greenbelt." *Rocky Mountain News*, January 14, 1973.

Gillen, Gary. "When Coal was King." Boulder *Daily Camera*, March 24, 1974.

———. "Queen Mary is Lifelong Friend to Erie." Boulder *Daily Camera* May 9, 1974.

Gits, Vickey. "Coal Miners' Memories." Boulder *Daily Camera*, April 11, 1976.

Leach, John. "Marshall Coal Mine Still Burning after 40 Years." Boulder *Daily Camera*, April 4, 1975.

Malmsbury, Todd. "Coal—Our Other 'Black Gold' Reserve." Boulder *Daily Camera*, January 25, 1976.

———. "Death Marked Mine Conflicts of the Past." Boulder *Daily Camera*, March 17, 1978.

Sampson, Joanna. "Coal Mine Tipple Gives Way to Years." Boulder *Daily Camera*, May 4, 1975.

Smith, Phyllis. "She Could Stop a Row or Riot." *Colorado Daily*, February 16, 1983.

———. "Coal Miners Strike, 1910-1914." *Colorado Daily*, April 17, 1983.

Taylor, Dick. "Coal Burns in Abandoned Marshall Mines." Boulder *Daily Camera*, July 10, 1956.

INDEX

NAMES OF MINES AND APPROXIMATE DATES OF CLOSING

1. Premier, 1944
2. Highview, 1944; Rosser, 1929; New Ross, 1940
3. El Dorado, 1939
4. Pine Cliff, 1942; Rosser, No. 7, 1907
5. Lewis, 1943; Lewis No. 2, 194?; Marshall, 1947
6. Eagle (old), 1936
7. Sunrise, 1936
8. Red Ash, 1952; Red Ash No. 2, 1937
9. Fox, 1936; Fox No. 2, 1936
10. Tropic, 1936
11. Crackerjack No. 2 (Cuba), 1958
12. Pittsburg, 1952; Ross, 1939
13. Crackerjack (old), 1958
14. Cook, 1928
15. Gorham, 1930; Gorham No. 2, 1935
16. Monarch No. 1, 1918; Mitchell, 1920
17. Crown (old), 1937
18. Black Diamond, 1931
19. Big Five, 1942; Crackerjack, 1958
20. Fox (old), 1923
21. Blue Goose, 1924
22. Allen Bond, 1895
23. Cactus, 1936; Glo-Coal (?); York, 1947
24. South Gorham, 1949
25. New Gorham, 1955
26. Pluto, 1951; Pluto No. 2, 1951
27. New Crown, 1955
28. Champion, 1927; Sunland, 1922: Matchless, 1927
29. Fireside, 1944
30. Industrial, 1945
31. Enterprise, 1898
32. Acme, 1928
33. Leader, 1899
34. Ajax, 1892
35. Centennial (old), 1952
36. Brooks (Nonpareil), 1925
37. Nonpareil, 1924
38. Monarch No. 2, 1947
39. Sunnyside, 1921
40. Imperial, 1898; Northern, 1938; Louisville, 1888
41. Rex No. 2, 1915
42. Caledonia, 1899
43. Rex, 1897
44. Hecla No. 1, 1920; Hecla No. 2, 1897
45. Electric, 1918
46. Big Lake, 1916; Garibaldi, 1916
47. Regal, 1947
48. Centennial (new), 1952
49. Vulcan, 1937
50. Hiway, 1954; Eversman, 1930
51. Liley, 1948
52. Paramount, 1939
53. Black Diamond No. 2, 1956
54. Strathmore, 1919
55. Cambro, 1928; Canon, 1890
56. Senator, 1913; Willoughby, 1908
57. Otis, 1898
58. Simpson, Spencer, 1890
59. Excelsior, 1899
60. Gladstone, 1906
61. Mitchell, 1918
62. Capitol, 1926
63. Standard, 1937
64. Caryl, Pallot, 1900
65. Vaughn, 1906
66. Baker, 1896
67. Blue Ribbon (Blue Goose), 1906
68. Haywood, 1906; Irvington, 1908
69. Mile High, 1918
70. Cleveland, 1895
71. McGregor, 1894
72. Marfel, 1904
73. Garfield No. 1, 1897
74. Stewart, Lister, 1890
75. Tynon, 1904
76. Chase, 1895
77. Northrup, 1884
78. Longs Peak, 1901
79. Jackson, 1890
80. Canfield, 1927(?)
81. Star (old), 1920
82. Unknown
83. Unknown
84. Arrow, 1935; Crescent, 1935; Star (new), 1935
85. Parkdale, 1916
86. Baseline, 1942; Lincoln (old), 1896, 1942; Park, Monroe, 1932; Washington, 1915
87. Lehigh, 1910
88. Mitchell, 1891
89. Garfield No. 2, 1905
90. Columbine, 1946
91. State, Andrew, 1917
92. Boulder Valley, 1969
93. Northwestern (Boulder Valley), 1907
94. Boulder Valley (old), 1893
95. Unknown
96. Reliance, 1907
97. Clayton, 1942
98. Morrison, 1953
99. Eagle
100. Imperial, 1972
101. Puritan, 1939
102. Shamrock, 1956
103. Munroe (Baseline), 1942
104. Erie Strip, 1953
105. Unknown
106. Whitehouse, 1907
107. Ideal, 1916
108. Eureka, 1926
109. Washington, 1967
110. Washington (old), 1911
111. Lincoln
112. Hingley-Morgan, (?)
113. Graden, 1955
114. Boulder Valley No. 3, 1954
115. Sterling, 1965
116. Witherbee, Silver State 1937
117. Peerless, 1919
118. Evans, 1942
119. Baum, Golden Ash, 1914
120. Frederick, 1929
121. Warwick, 1912
122. Firestone, 1920
123. Grant, 1939
124. Russell, 1947
125. Emerson, 1904
126. St. Vrain, 1892
127. McKissic, 1907
128. Davies, 1906
129. Welch, 1896(?)
130. Shannahan, 1900

Base from U.S. Geological Survey, 1:24,000: Eastlake, 1965, Erie, 1950-67, Frederick, 1950 (69PR), Gowanda, 1949 (69PR), Lafayette, 1965, and Louisville, 1965

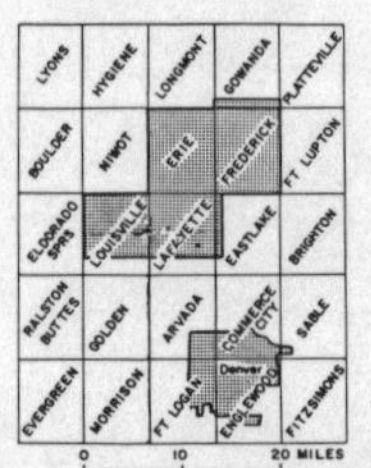

INDEX SHOWING LOCATION OF AREA IN RELATION TO DENVER AND ADJOINING QUADRANGLES

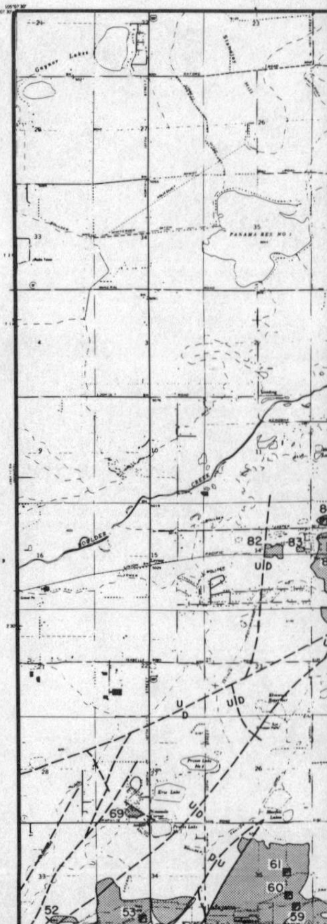
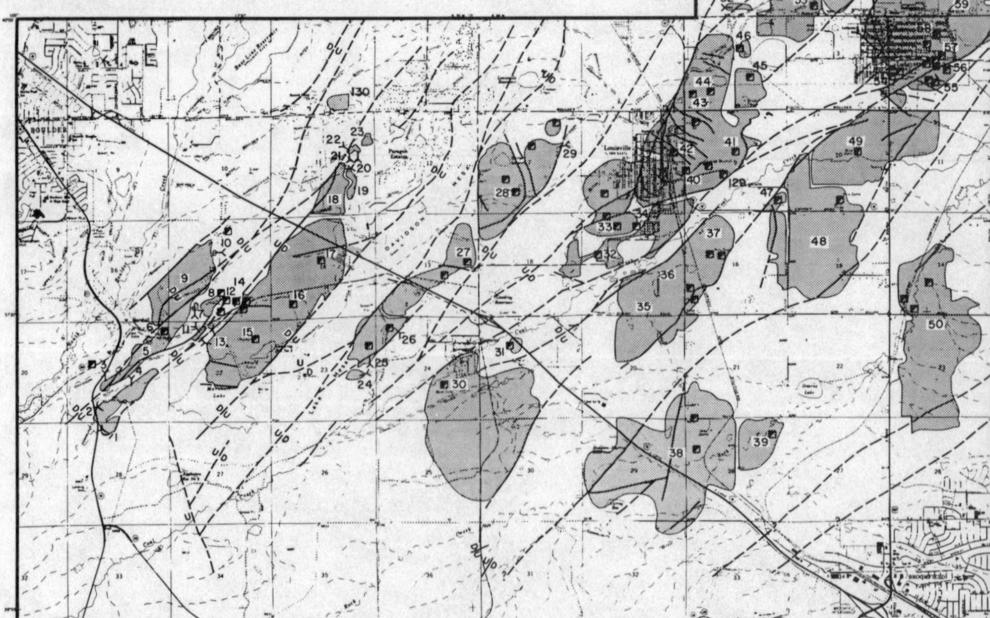